Interdisciplinarity, Creativity, and Learning

Mathematics with Literature, Paradoxes, History, Technology, and Modeling

T0338429

Interdisciplinarity, Creativity, and Learning

Mathematics with Literature, Paradoxes, History, Technology, and Modeling

edited by

Bharath Sriraman
The University of Montana

Viktor Freiman
University of Moncton

Nicole Lirette-Pitre
University of Moncton

INFORMATION AGE PUBLISHING, INC.
Charlotte, NC • www.infoagepub.com

Library of Congress Cataloging-in-Publication Data

Interdisciplinarity, creativity, and learning mathematics with literature, paradoxes, history, technology, and modeling / edited by Bharath Sriraman, Viktor Freiman, Nicole Lirette-Pitre.

    p. cm. – (The Montana mathematics enthusiast, monograph series in mathematics education)
 Includes bibliographical references.
 ISBN 978-1-60752-101-3 (pbk.) – ISBN 978-1-60752-102-0 (hardcover)
1. Mathematics–Study and teaching. 2. Interdisciplinary approach in education. I. Sriraman, Bharath. II. Freiman, Viktor. III. Lirette-Pitre, Nicole.
 QA11.2.I66 2009
 510.71–dc22

                                        2009011496

This monograph is made possible by support from the Canadian Social Sciences and Humanities Research Council Developmental Grant entitled Interdisciplinary networks for better education in mathematics, science, and arts.

Printed in the United States of America

Interdisciplinarity is increasingly viewed as a necessary ingredient in the training of future oriented 21st century disciplines that rely on both analytic and synthetic abilities across disciplines. Nearly every curricular document or vision statement of schools and universities include a call for promoting creativity in students. Yet the construct of creativity and giftedness across disciplines remains elusive in the sense that the prototypical examples of such work come from eminent scientists, artists and mathematicians, and little if any work has been conducted with non-eminent individuals. This monograph is an attempt to fill this gap by putting forth the view that interdisciplinarity and creativity are related constructs, and that the cultivation of domain general creativity is possible. Mathematics has historically been anchored to numerous disciplines like theology, natural philosophy, culture and art, allowing for a flexibility of thought that is difficult to cultivate in other disciplines. In this monograph, the numerous chapters from Australia, U.S.A., Canada, Cyprus, Denmark and Japan provide a compelling illustration of the intricate connection of mathematics with literature, paradoxes, history, technology and modeling, thus serving as a conduit for interdisciplinarity, creativity and learning to occur.

CONTENTS

SECTION I

INTERDISCIPLINARITY IN MATHEMATICS AND LITERATURE

SECTION II

MATHEMATICS AND PARADOXES

SECTION III

GEOMETRY AND HISTORY

SECTION IV

INTERDISCIPLINARITY AND MODELING

SECTION V

TECHNOLOGY AND THE NET GENERATION

SECTION I

INTERDISCIPLINARITY IN MATHEMATICS
AND LITERATURE

CHAPTER 1

THE INTERDISCIPLINARY NATURE OF INDUCTIVE PROCESSES IN FORMING GENERALIZATIONS

Bharath Sriraman
The University of Montana

Harry Adrian
Ottawa Township High School, IL

ABSTRACT

Human thought has the capacity to generalize from specific experiences leading to new, more abstract concepts. Generalizations are the end result of an inductive process that begins with the identification of similarities in seemingly disparate situations. It is the existence of such generalizations that makes it possible for us to understand each other and the world around us. It is pedagogically weak to present generalizations to students and expect them to know how and when to apply them. On the other hand if students experience the inductive process in classrooms and discover generalizations, they are likely to remember and replicate this process when tackling other problems. The authors illustrate the interdisciplinary nature of the inductive

Interdisciplinarity, Creativity, and Learning, pages 3–12
Copyright © 2009 by Information Age Publishing

3

process by applying it to two seemingly different domains, namely English literature and Mathematics. In this paper the inductive process is applied to four short stories and four problem-solving situations in mathematics, which results in arriving at plausible generalities that characterize the stories and the problems. A conceptual model that illustrates how inductive processes facilitate generalizations in the classroom is presented.

Two roads diverged in a wood, and I—I took the one less traveled by, and that has made all the difference.

—Robert Frost

INTRODUCTION

There is an old adage which states that giving a hungry man a fish feeds his hunger temporarily whereas teaching him how to fish would feed him forever. This old adage has relevance in the context of teaching and learning in higher education. When students are confronted with a difficult problem, they inevitably ask the teacher for help. The teacher usually solves the problem for the student using methods from their repertoire and the student receives instant gratification for that particular problem. The next time the student is confronted with a difficult problem, the student instinctively approaches the teacher, receives the solution, and this pattern continues. How can we expect students to ever think for themselves if we do the thinking for them?

For example in mathematics, it is often the case that students are concerned about the correctness of their solutions. A common scenario in many classrooms in the United States is the routine reading out of the "answers" to homework problems, followed by the teacher demonstrating the solutions to one or more problems that students missed. There is very little effort put forth in discussing incorrect solutions and the reasons why the strategy employed by a student did not yield the desired solution. Students are expected to apply a priori procedures covered in the classroom and are not challenged to use either novelty or originality in their solution attempts. Similar scenarios are everyday occurrences in classrooms across the United States and India not just in mathematics but in other domains as well. Another example would be in the domain of literature, where teachers often lead students into a priori interpretations, and thus preventing the students from developing an original point of view. These statements are based on the authors schooling, teaching and collaborative research experiences in the United States and India.

The *National Council of Teachers of English* (NCTE) in the U.S calls for teachers to subscribe to certain guiding principles such as understanding the role of language in learning, understanding the power of story as the basic unit of mind, understanding the power of literature to reawaken the imagination and understanding the power of writing as a tool for thinking across the curriculum (Harste,1999). These goals are praiseworthy and transcend classroom cultures across the world. In the same vein, the *National Council of Teachers of Mathematics* (NCTM, 2000) calls for instructional programs that emphasize problem solving; classrooms that foster creativity and a climate of investigation that helps students develop productive mathematical dispositions. It is clearly the case that no matter what the domain of knowledge is, teachers would like students to develop their own thoughts and powers of reasoning, with the hope of producing independent thinking citizens who can carve an original niche in society.

This implies that the teacher should be the "gadfly" who questions rather than preaches. The teacher's responsibility is to teach the student how to think rather than what to think. Teachers need to impart to the students that "understanding" is a synthesis of knowledge and insight brought about by discovery and arriving at the inductive leap. Yet many students are doing just the opposite, which is deducing how to play the teacher's game.

So how does one go about creating a classroom that fosters original thought? How can we lead students into discovering and formulating generalizations? These are the questions of exploration in this paper.

BACKGROUND

Two patterns of reasoning are commonly acknowledged across cultures. They are deductive and inductive methods of reasoning. Deductive reasoning starts from the generality and arrives at the specific. In mathematics there are numerous general theorems that yield diverse particular examples. In the realm of science one comes across theories which in turn can be narrowed down to hypothesis. These hypotheses can then be subject to test, through observations, which either lend validity to the theory or observations that run contrary to the theory. In literature deduction becomes almost algorithmic when interpretation depends on teacher's perspective.

On the other hand, inductive reasoning works in the opposite direction. It begins with particular cases or observations within which patterns are detectable. This in turn results in the formulation of generalities and theories. Inductive processes begin with the identification of similarities, commonalties and/or patterns in seemingly disparate situations. These processes are naturally kindled by contextual, social and cultural interaction that takes place in the classroom. The creation of mathematics entails "not making useless combinations but making those which were useful and which are only a small minority" (Poincaré, 1948). In other words the act of mathematical creation was an inductive zig-zag path of trial and error which began with the construction of examples, within which plausible patterns were detectable, which in turn led to the formulation of theorems (Hardy,1956; Lampert,1990; Polya,1954). In literature when interpretation depends on teacher's perspective, an inductive approach is that in which the teacher's thoughts are "wedded" to the student's thoughts arriving at values codified by culture. Thus, questioning becomes a crucial part of the inductive process. Leo Tolstoy praises the value of questioning as follows:

> The peasants say that a cold wind blows in late spring because oaks are budding, and really every spring cold winds do blow when the oak is budding. But I do not know what causes the cold winds to blow when the oak buds unfold. I cannot agree with the peasants that the unfolding of the oak buds is the cause of the cold wind, for the force of the wind is beyond the influence of the buds. I see only a coincidence of occurrences such as happens with all phenomena of life, and I see that however much and however carefully I observe...the oak, I do not discover the cause of the...winds of spring. To do that I must change my point of view and study the laws of the wind. (Leo Tolstoy in *War and Peace*)

One concern of the authors was whether or not inductive thinking was a cross-cultural phenomenon? A careful perusal of recent research literature revealed the following. In one study (Vijver,1991), tests of inductive thinking which consisted of items that called for classification, rule generation and rule verification, were administered to 532 Dutch, 877 Turkish, 141 Turkish-Dutch and 704 Zambian students aged 10–16 years. It is noteworthy that the results indicated that inductive thinking was universal in these diverse populations, providing evidence that inductive processes are similar across cultures. In another recent longitudinal study in Germany (Hamers, de Koning, Sijtsma, 1998), teachers were asked to implement an inductive approach in third grade classrooms with the hope of enhancing inductive reasoning abilities of the students. The researchers found that students in these experimental classrooms outperformed the control group students in tests administered immediately after the experiment as well as on tests administered almost four months later. This provides additional evidence to the pedagogical value of nurturing an inductive approach to learning in the classroom.

METHODOLOGY

The authors will now illustrate the pedagogical power of the inductive process by applying it to two seemingly different content areas, namely English literature and Mathematics. The inductive process will be applied to four short stories and four problem-solving situations in mathematics, which will result in arriving at plausible generalities that characterize the stories and the problems.

Application of the Inductive Process in the Domain of English Literature

Consider an English literature course, which consists of reading a collection of short stories from different authors. The teacher would ideally like students to discover generalizations (or common threads) that connect the stories that are seemingly very different from each other. For illustrative purposes consider the following stories.

TABLE 1.1 Four Diverse Short Stories

Story 1: *The Garden Party* by Katherine Mansfield.
Story 2: *The Three-Day Blow* by Ernest Hemingway.
Story 3: *The Standard of Living* by Dorothy Parker.
Story 4: *The Necklace* by Guy de Maupassant.

The Garden Party (Story 1) is a short story about the perception that the idle rich have of the working class. This story vividly depicts the totally different value systems held by the very rich and the poorest of the poor. These value systems clash although they are understandable within each perspective. One sees a gradual change in the sensibilities of the naive main character, Laura, as she is exposed to the thoughts of her siblings (Meg and Laurie) and her parents (The Sheridians). A garden party with all the festivities is contrasted with the death of a poor person who lives on the edge of the Sheridian's land. Laura, conscience stricken desires to cancel the party out of respect but her parents would not. The contrast of life-styles is never so evident as when Laura takes the left over sandwiches to the funeral of the poor person who was killed. The frustration and inability of Laura to understand life is never more obvious as when she stammers at the end of the story "Isn't life . . . .?" And Laurie answers "Isn't it darling?"

The Three-Day Blow by Ernest Hemingway (Story 2) is the story of two men with different ideas about their relationships with women. Nick is vaguely aware of his pain on losing Majorie. Bill has a cavalier attitude, which seems to translate into seeing women more as objects to be used rather than as another person. He also has a low opinion of marriage. *The Three-Day Blow* is about the death of love. It is also about using others as objects for use rather than as another person to share with or sympathize about similar problems.

The Standard of Living by Dorothy Parker (Story 3) is the story of two young ladies trapped by their profession and lifestyle. They are defined as two dreamers who lack the ability to move beyond their environment. They are limited in terms of their social inability defined by the author as "conspicuous, cheap, an charming."

Finally, *The Necklace* by Guy de Maupassant (Story 4) is the story of a young woman who is trying to "keep up with the Jones." Instead of being content with her new ballroom gown, Madame Loisel insists on wearing an expensive necklace to accentuate her status and beauty. The following chain of events result in the necklace being lost, and Madame Loisel toiling for ten years to pay off the debt that arose from having to replace the necklace. The story ends with the irony that the lost necklace was a fake.

It is fairly evident that teachers have to exercise a certain amount of choice in the selection of stories that will foster an inductive approach. The teacher can initiate the inductive process by a series of anticipatory questions formulated to get students to reflect about the value systems that play a role in each of the four stories. After reading a short story discussion follows and the teacher can use the Socratic method of questioning to force students to analyze the characters actions and motivations demonstrated in each of the stores. This leads students into identifying commonalties in the stories and formulating generalizations. One generalization could be the effect that siblings and parents have upon other members of the family. Usually family life tends to form value systems which in a sense "hardwire" those members to a particular perspective. Another generalization might be how each person in a friendship or family relationship has along with common beliefs divergent ideas about life and how to live it. A final generalization might be a realization that one must be a critical thinker, i.e., a person who thinks about his/her thinking while h/she is thinking to recognize bias brought about by culture, family, religion, education, peers, etc. This allows the student to understand the motivation each character has to behave and/or feel as revealed in the stories. The authors will now apply the inductive process to the less subjective domain of mathematics in order to illustrate its applicability across disciplines.

Application of the Inductive Process in the Domain of Mathematics

Consider a mathematics classroom within which the teacher wants students to become adept at problem-solving as well as being able to reason and formulate generalizations. The teacher presents the following four problem solving situations (Table 1.2) as recreational problems to the students over the course of several weeks.

TABLE 1.2   Four Diverse Problem-Solving Situations

Situation1 (The Soda Problem)

The soda menu of Newton's Bistro has six choices for sodas, namely Cola, Diet Cola, Orange Soda, Ginger Ale, Root Beer, and Pink Lemonade.

How many students would be required to place soda orders, one soda per student, in order to insure that at least one of the six listed sodas would be ordered by at least two students?

Situation 2: (The Aspirin Problem)

A person takes at least one aspirin a day for 30 days. Suppose he takes 45 aspirin altogether. Is it possible that in some sequence of consecutive days he takes exactly 14 aspirin? Justify your solution. Prove your assertion.

Situation 3: (The Number Sum Problem)

Choose a set S of ten positive integers smaller than 100.

For example choose the set S = {5, 14, 44, 16, 29, 53, 46, 61, 89, 80}

There are two completely different selections from S that have the same sum.

For example, in the set S, one can first select 5, 14 and 61, and then select 5, 29,and 46. Notice that they both add up to 80. (5 + 14 + 61 = 80; 5 + 29 + 46 = 80).

One could also first select 14, 29, 46 and then select 89. Notice that they both add up to 89 (14 + 29 + 46 = 89; and 89 = 89).

No matter how one chooses a set of *ten positive integers* smaller than 100, there will always be two completely different selections that have the same sum.

Why does this happen? Prove that this will always happen.

Situation 4: The Party Problem

There are 50 people at a party. Some of them are acquainted with each other, some not. Prove that there are two persons in the room who have an equal number of acquaintances.

To the reader it would seem that the four situations stated above have absolutely nothing in common. However pursuing solutions to these four problems and reflecting on the structure of the solutions reveal the common thread that binds them.

The Soda problem (Situation 1) has the very obvious solution that seven students would be required to place soda orders, since the worst case scenario is each of the first six students order a different drink, thus forcing the seventh student to order a drink that has been previously ordered.

The Aspirin problem (Situation 2) is commonly resolved by assuming that the person takes at least one aspirin pill a day. Therefore in thirty days, the person would have consumed exactly thirty aspirin pills, thus leaving a surplus of fifteen pills, which the person can randomly take in the thirty-day cycle. Let a_i be the total number of aspirin consumed up to and including the ith day, for $i = 1, \ldots, 30$. Combine these with the numbers $a_1 + 14, \ldots, a_{30} + 14$, providing 60 numbers, all positive and less or equal to $45 + 14 = 59$. Hence two of these 60 numbers are identical. Since all a_i's and, hence, $(a_i + 14)$'s are distinct (at least one aspirin a day consumed), then $a_j = a_i + 14$, for some $i < j$. Thus, on days $i + 1$ to j, the person consumes exactly 14 aspirin.

The Number Sum problem (Situation 3) can be solved as follows. There are $2^{10} = 1{,}024$ subsets of the 10 integers, but there can be only 901 possible sums, the number of integers between the minimum and maximum sums. With more subsets than possible sums, there must exist at least one sum that corresponds to at least two subsets. Hence there are always two completely different selections that yield the same sum.

Finally, the Party problem (Situation 4) can be resolved as follows. If there is a person in the room who has no acquaintances at all then each of the other persons in the room may have either 1, or 2, or 3,..., or 48 acquaintances, or do not have acquaintances at all. Therefore we have 49 "holes" numbered 0, 1, 2, 3,..., 48, and have to distribute between them 50 people. Next, assume that every person in the room has an acquaintance. Again, we have 49 holes 1, 2, 3,..., 49 and 50 people. Thus two people will be forced to have the same number of acquaintances.

The teacher can initiate the inductive process by transmitting practices that call for reflection on the structure of the solutions. The underlying structure of the solutions reveals that there are always two quantities that play a role in each of the four problems, and the problem is solvable because one of these quantities is larger than the other one, thus "making" things happen. Reflecting on the structure of the solutions leads one to the discovery of the powerful generalization that characterizes these four problems, namely the pigeonhole (or Dirichlet) principle. If there are "m" pigeons are put in "n" pigeonholes and $m > n$, then there is a hole with more than one pigeon. In other words if there are more pigeons than pigeonholes, then this forces some holes to have more than one pigeon. This generality is the result of inducing commonalties from the four particular problems over a period of time.

IMPLICATIONS

Having illustrated the usefulness of the inductive process, the authors will now suggest a conceptual model (Figure 1.1) within which such a pedagogical approach could be put into practice. This model takes into account the social and cultural dimensions of the inductive process, instead of simply viewing it as an individualistic psychological process. This approach takes into account the interaction between an individual student, the classroom, the teacher, and the culture.

This model (Figure 1.1) suggests that generalizations are the result of the interaction between the individual, the classroom, the teacher and the

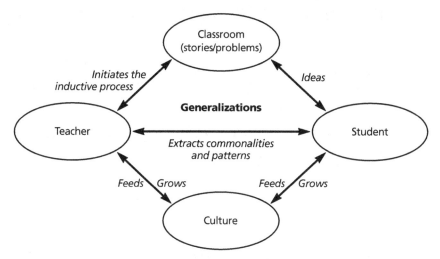

Figure 1.1   Inductive processes that facilitate generalizations in the classroom.

culture. One cannot simply focus on the individual aspects of the inductive process because the individual operates within the culture of the classroom and is influenced by the practices transmitted by the teacher and other students.

Thus generalizations occur when the teacher initiates the inductive process in the classroom and transmits this through classroom practices, such as initiating a discussion on observable patterns in stories/problem. This generates a flow of ideas between the student, the teacher and the story/problem. Culture also influences the generation of ideas. The student then extracts commonalties and patterns and this flows between the student and the teacher. This interaction results in the formation of generalizations. The authors contend that by consistently cultivating rules and practices that initiate the inductive process, students will produce novel variations in the commonalties and patterns presented in stories/problems, thereby resulting in generalizations. Generalizations are subject to the test of time, after which they may become a part of the culture.

REFERENCES

Hamers, J., de Koning, M., Sijtsma, K. (1998). Inductive reasoning in third grade: Intervention, promises and constraints. *Contemporary Educational Psychology,* *23*(2), 132–148.

Hardy, G. H. (1956). A mathematician's apology. In J. R. Newman (Ed.), *The world of mathematics*. New York: Simon & Schuster.

Harste, J. C. (1999) *Re-imagining the possibilities of NCTE.* Speech at NCTE Affiliate Roundtable Breakfast. http://www.ncte.org/issues/coreintro.shtml.

Lampert, M. (1990). When the problem is not the question and the solution is not the answer: Mathematical knowing and teaching. *American Educational Research Journal, 27,* 29–63.

National Council of Teachers of Mathematics. (2000). *Principles and standards for school mathematics.* Reston, VA: Author.

Poincaré, H. (1948). *Science and method.* New York: Dover.

Polya, G. (1954). *Mathematics and plausible reasoning: Induction and analogy in mathematics* (Vol. II). Princeton, NJ: Princeton University Press.

Vijver, V. (1991). Inductive thinking across cultures. *Dissertation Abstracts International.* Vol. 52-04, Section C, 0674.

ACKNOWLEDGMENT

Reprint of Sriraman, B., & Adrian, H. (2004). The pedagogical value and interdisciplinary nature of inductive processes in forming generalizations. *Interchange: A Quarterly Review of Education, 35*(4), 407–422. Reprinted with permission from Interchange. © Bharath Sriraman

CHAPTER 2

THE EXISTENTIAL VOID IN LEARNING

Juxtaposing Mathematics and Literature

Bharath Sriraman
The University of Montana

Harry Adrian
Ottawa Township High School, IL

ABSTRACT

The term "existential" is normally used in the context of the human search to give meaning to existence. This metaphor could be used in the context of learning. A student, whose learning experiences in school lacks moral or ethical guidelines which does not have the power to give meaning to her world, experiences the existential void. Even the brightest students experience this void. On the other hand many are not aware of this because they are caught up in their world of rock and roll. The teachers role is to create learning experiences that help students fill the void. Mathematics has symmetry or a totality that blends the parts to the whole. It is a continuum as opposed to pieces of a disjointed puzzle. Mathematics can allow students to experience the exhilaration of discovery as well as see its connections to areas in the arts, business, and sciences. Similarly, the study of literature through the prism

Interdisciplinarity, Creativity, and Learning, pages 13–29

of critical thinking can also allow the student to experience its cohesiveness to life. Literature can be practical, inspirational, appealing, stimulating and educational if approached with this purpose in mind.

Ideally the goal of learning is to extend vision, to broaden perspective, and to bring out coherence and unity among the disciplines. The general goal of this paper is to demonstrate how education can fill the existential void felt by students, and to give education purpose and viability. In particular, the authors will use one classical mathematical problem and one contemporary literary work to show how teachers can fill the existential void in learning.

INTRODUCTION

The term *existential* may be used in three ways. To refer to (1) existence itself; (2) the meaning of existence; and (3) the striving to find a concrete meaning in personal existence, that is to say, the will to meaning (Frankl, 2000). Applying the third way "the will to meaning" to education is the question for exploration in this chapter.

Ideally the goal of learning is to extend vision, to broaden perspective, and to bring out coherence and unity among the disciplines. A student whose learning experiences in school involve the mindless regurgitation of facts, figures and formulae, which lack meaning nor add beauty to her world, experiences the existential void. The standard rationale that education means fulfillment begs the question: how can education fill the existential void felt by students? The answer may seem simplistic and/or idealistic but the onus should be placed on the teacher. It is the teachers role to create learning experiences that help students fill the void and to give education purpose and viability. The authors will draw on their experiences as teachers in the field of literature and mathematics to illustrate how a subject matter can be used to initiate the search for meaning.

The study of literature can be practical, inspirational, appealing, stimulating and educational if approached through critical thinking, which in turn can allow the student to experience its connections to life. Similarly mathematics can allow students to experience the cohesiveness of an idea from the classroom, to numerous applications in the various sciences, as well as connections to underlying patterns in nature, art and music. For example, a particular branch of mathematics called number theory has profound applications to modern day life, in the guise of cryptography and computer science.

USING CONJECTURE–PROOF–REFUTATION
TO FILL THE EXISTENTIAL VOID

Most high school students in the United States view mathematics as consisting of immutable truths. For example, in geometry students believe that remembering theorems, facts and rules is the primary goal and very few understand the relationship between definitions, postulates, axioms and theorems (Fawcett, 1938; Senk, 1985, Usiskin, 1987). Within mathematical philosophy, mathematical knowledge is espoused as either absolutist or fallibilist. The absolutist view suggests that mathematical knowledge consists of certain and unchallengable truths (Ernest, 1991). The fallibilist view of knowledge asserts that mathematical knowledge is "fallible and corrigible, and can never be regarded as beyond revision and correction" (Ernest, 1991, p. 18). The philosopher Lakatos (1976) viewed mathematics as an ongoing process of conjecture, proof, and refutation, and this is the view that the authors subscribe to. In other words the authors subscribe to a fallibilist epistemology and favor instructional practices, which are more open to the influence of students.

In the United States in most traditional high school curricula, students do not experience the process of establishing a mathematical truth until they encounter geometry. In fact when students first study geometry, they are immediately introduced to a deductive method of proof, which deprives them of the process of discovery of the mathematical truth. Inductive reasoning plays a major role in the real world. However, it is only in mathematics that one comes across the notion of a proof, whose sole purpose is to establish the truth (or falsity) of a given statement. The reasoning that one normally comes across in many mathematics textbooks is crisp and deductive, with one statement flowing from another until the desired outcome is reached. An artificially reconstructed logical proof conveys little or no insight to the student about the processes and the motivations for constructing the proof.

The last decade has shown some change to this traditional method of studying geometry with the introduction of dynamic computer software such as the Geometer's sketchpad and Cabri Geometry. Although such software has great capabilities, most of the anticipatory questions would have to be carefully designed by teachers if they wish to guide students into the discovery of a mathematical truth. However many rural and less affluent school districts lack the technological and human resources that would enable students to study geometry using an inductive approach on a medium such as the Geometer's Sketchpad.

Aside from geometry, students in their study of two years of high school algebra encounter very little experience with proof. In fact most traditional algebra curricula takes the properties of the real number system as a priori truths and focuses on analytic representations of geometric figures on the Cartesian co-ordinate system with subsequent manipulations of polynomial equations and inequalities. A substantial amount of time is also spent on solving contrived "word problems" under the guise of "applications" in order to somehow justify the topics that have been covered. There is very little room for 'play' or exploration within such an approach. This approach also does not convey to the student that mathematics is a process of ongoing conjecture, proof and refutation (Lakatos, 1976).

USING A CLASSICAL DIOPHANTINE PROBLEM TO INITIATE CONJECTURE–PROOF–REFUTATION

The Greek mathematician Diophantus (200 AD) is renowned for his work on solving equations with rational number solutions. Number theorists relish tackling diophantine equations with integer solutions. The beauty of many diophantine equations lies in the fact that they are easy to understand, yet very difficult to solve. Fermat's Last Theorem is a notorious example to illustrate this point. Elementary number theoretic concepts such as prime numbers, and tests for divisibility are introduced in most middle school curricula. However at the high school level, the curriculum offers students very little opportunity to tackle number theoretic problems. For example, many questions in number theory may be posed as diophantine equations— equations to be solved in integers. Catalan's conjecture was whether 8 and 9 were the only consecutive powers? This conjecture asks for the solution to $x^a - y^b = 1$ in integers. The Four Squares Theorem states that every natural number is the sum of four integer squares. In other words, it asserts that $x^2 + y^2 + z^2 + w^2 = n$ is solvable for all n. But the attempt to solve these equations requires rather powerful tools from elsewhere in mathematics to shed light on the structure of the problem (Oystein, 1988).

This led the author to incorporate diophantine equations as the teacher of a beginning algebra course. Students in this course were introduced to elementary diophantine equations under the guise of recreational journal problems. The problem chosen for investigation was the classic n-tuple diophantine problem posed by Diophantus himself. Simply put, a diophantine n-tuple is a set of n positive integers such that the product of any two is one less than a square integer. It was the authors' hope that a very elementary version of the problem would kindle student interest and eventually result in an attempt to tackle the as yet unsolved 5-tuple problem in integers. Does there exist a diophantine 5-tuple? Many mathematicians propose this problem as the successor to Fermat's last theorem.

The author initiated this problem by simply mentioning in class off hand the 3-tuple problem, if one considers the integers 1, 3, and 8, then it is always the case that the product of any two is always one less than a perfect square. Indeed $1 \times 3 = 2^2 - 1$; $1 \times 8 = 3^2 - 1$; and $3 \times 8 = 5^2 - 1$. This remark led students to wonder if other such 3-tuples existed? This problem was then assigned as a recreational journal problem. This simple question eventually led to an investigation of the unsolved 5-tuple problem over the course of the school year (see Figure 2.1).

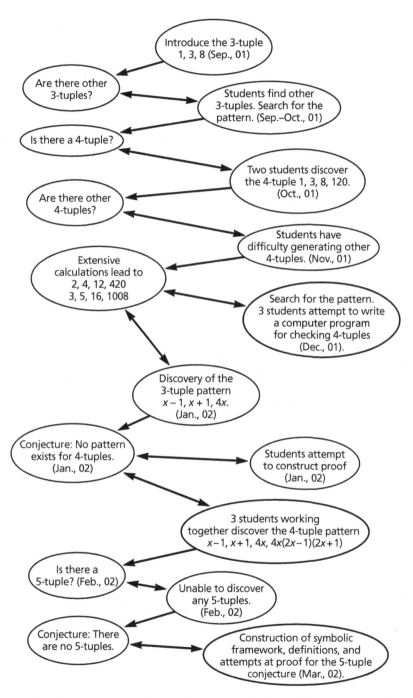

Figure 2.1   Conjecture, proof, and refutation in the 5-tuple Diophantine problem.

As indicated in Figure 2.1, the students attempted to solve the open 5-tuple diophantine problem being immersed in the process of conjecture, proof and refutation. Once a week, the teacher utilized one class period to allow students to present their work to the other students. The teacher acted as a moderator in this process to facilitate the discussion. Many of the students were surprised at the difficulty of solving these seemingly easy problems. The mathematical experiences of several students also led them into writing algorithms and computer programs to check for integer solutions. Usually when other students in the class became convinced that multiple integer solutions to a particular problem existed, they formed conjectures on what the underlying pattern was and tried to express them symbolically. This in turn allowed the teacher to use student conjectures as a starting point to initiate the process of mathematically proving or disproving a conjecture. It is important that the reader note that the progress of the problem depended completely on the "will" of the students. There were some weeks when students did not attempt the problem, especially before or after school breaks and during examination weeks. When this occurred, the author continued with the regular algebra curriculum. It was crucial that students initiated this process of conjecture, proof and refutation out of their need to resolve the unimagined difficulties that arose from a seemingly easy problem. It was noteworthy that all thirteen students in this class willingly engaged in trying to solve one of the unresolved conjectures of our time over a seven month time period through the process of conjecture, proof and refutation. The interested reader can find the actual mathematics created by the students in Appendix A.

USING CRITICAL THINKING IN LITERATURE TO FILL THE EXISTENTIAL VOID

The process of conjecture, refutation and proof does have a counterpart in the realm of literature. The process of critical thinking could be viewed as the use of reasoning in the pursuit of "truth." Critical thinking makes implicit use of logic in order to draw inferences and/or make comparisons. Critical thinking enables the student to understand the cultural and instructional influences on accepted thought. Because of this "second sight"-so to speak, the student can adjust or trick the mind into a new view of the issue. True critical thinking is a matter of adjustment to tune the process, so that bias no longer controls thought or action.

In the previous section, the authors used the diophantine n-tuple problem, in order to kindle the pursuit of a mathematical truth by the process of conjecture, proof and refutation. In a similar spirit, the authors will now use a contemporary novel in order to demonstrate how a simple story can be used to initiate students into the process of critical thinking and into making inferences on "truths" about society and life. The "truths" that we infer are naturally influenced by our social, economic, cultural, religious backgrounds, and value systems.

The English philosopher Francis Bacon warned about blind observance to so called "truths" in his *Novum Organum*, using the metaphor of "the idols of the mind" (Bacon, 1994). There are four classes of idols which beset men's mind, namely, the idols of tribe, the idols of cave, the idols of marketplace, and idols of theater. Titus (1994) describes Francis Bacon's "idols" metaphor in the following words.

> Bacon has given us a classic statement of the errors of thinking. These are first, the idols of the tribe. We are apt to recognize evidence and incidents favorable to our own side or group (tribe or nation). Second, there are the idols of the cave. We tend to see ourselves as the center of the world and to stress our own limited outlook. Third, the idols of the marketplace cause us to be influenced by the words and names with which we are familiar in everyday discourse. We are led astray by emotionally toned words-for example, in our society, such words as communist or liberal. Finally, the idols of theater arise from our attachment to parties, creeds, and cults. These fads, fashions, and schools of thought are like stage plays in the sense that they lead us into imaginary worlds; ultimately, the idols of theater lead us to biased conclusions. (Titus, 1994, p. 171)

Using a Contemporary Novel to Initiate Critical Thinking

Grisham's (2001) *Skipping Christmas* is a retelling of Charles Dickens' *A Christmas Story* with contemporary characters facing contemporary problems. Using life in suburbia as his backdrop, Grisham (2001) weaves his tale of Luther "Scrooge" Krank and his dream of skipping Christmas and all the baggage this holiday carries.

Luther is a character devoid apparently of any Christmas spirit. Tradition is abhorrent to him and his tirades against the idea of Christmas are a litany of all the negative clichés about the commercialization of Christmas. From tipsy office partygoers, thoughtless gifts, crass symbols, to dollars spent, Luther condemns Christmas to the "bah humbug" status of the typical Scrooge. He decides, and then he cons Nora, his wife, into going on a

Caribbean cruise starting on December 25. This is Luther's ruse to miss out on Christmas and all that goes with the season.

Grisham (2001) using all the usual traditions as his canvas then paints the hazardous tale of Luther and Nora's decision. He veers from his usual deep, dark sub-plots although sub-plots abound. These sub-plots involve traditional value systems, which have become part of the American canon or code. Family, love, co-operative spirits, concern and love of neighbor, respect for the beliefs of others, equality of birth, sharing and finally the rights of all of us, life, liberty and the pursuit of happiness are just some of these traditional values touched upon.

In a sense the novel seems to be an appeal to past values. In the world of Luther values have become passé signifying our modern world of "me-centric" existence. In some cases the youth of today have become deprived of an ethical inheritance. No thing, no person, no ideal, no code has come to fill the existential void. The events of September 11, 2001 jolted many Americans into remembrance of time past. That was then, this is now has been turned around by this horrendous event. Americans, especially the young do seek the values and security of the past. *Skipping Christmas* is Grisham's (2001) attempt to review and renew some of those values.

The literature teacher through discussion of this short novel has the opportunity to review the basic values of our democracy. The destruction of the twin towers of honesty and truth by the political and business leaders of the past few years has deprived our youth of a continuity of values. Thus, the deadly existential void has become apparent.

Using *Skipping Christmas* as a critical thinking didactic tool, literature teachers can provide opportunities for students to fill the existential void they may be experiencing. The role of the teacher in this process can be thought of as that of an eye specialist rather than that of a painter. A painter tries to convey to us a picture of the world as he sees it; the eye doctor tries to enable us to see the world as it really is. It is crucial that teachers not be propagandists or try to indoctrinate students. The teachers role through critical thought extends the visual field of the student, so that the whole spectrum of potential meaning becomes visible to him/her (Frankl, 2000). With this philosophy of interpretation in mind, students can be exposed to the novel. There are various strategies that teachers can use to kindle critical thinking in the students. The teacher can use critical thinking questions to set up a forum for discussion. For example, comparisons can be made to Dicken's timeless fable *A Christmas Carol.* Dicken's novel is about a character Scrooge and his all consuming love of money. The spirit of Christmas is a contrast to his selfishness and lack of love, joy and care for others. It is the story of the haunting of Ebenezer Scrooge by ghosts of Christmas past, present and yet to come. These ghosts drive Scrooge to reconsider his life and construct a new mode of living. *Skipping Christmas* is also about just such a character, Luther Krank. Even his last name "Krank" seems to depict his personality.

An English teacher can also use an anticipatory questionnaire using student answers for discussion in a manner akin to conjecture, proof and refutation. The anticipatory questions must be designed not to set a cause effect pattern of thought. This can range from particular questions to general views about the subject matter. One needs almost to upset the natural biased way of interpretation. What results then, is a personal pattern rather than an algorithmic type of answer. In literature various interpretations lead to critical thought actually arrived at by the student, rather than "I have found what you-the teacher are thinking." The student's thought as well as teacher's thought combined reaches a valued conclusion. This of course must be tested in the "market place of ideas." The authors will now present the results of an experiment in critical thinking using *Skipping Christmas* as a didactic tool. Eight high school seniors who read the novel were invited to discuss the book. The story was used to frame five general anticipatory questions (see Appendix B), which kindled a discussion of personal value systems and values in general. In other words, the book was simply a didactic tool to initiate critical thinking about traditions and value systems and used by the teacher to provide a framework for the discussion of the five questions.

The authors will now show the progression of thought as the novel is delineated by the use of the Socratic method of discovery. In an effort to show that a continuity of thought exists from question to answer to subsequent conclusion, student responses from the discussion are presented in the form of a schematic analogous to conjecture-proof-refutation (see Figures 2.2–2.5).

As the discussion schemes outlined in Figures 2.2–2.5 reveal, daily class discussion using anticipatory questions and the answers of students as guidelines or starting points is a powerful method of instruction This method allows the teacher to introduce historical and philosophical perspectives to moral questions. These perspectives combined with the student's views leads students to examine codes of behavior and biases. The teachers role is to create a classroom atmosphere that nurtures and kindles critical thinking in students, so that students begin to examine their biases and have the opportunity to discuss their value systems and their perspectives.

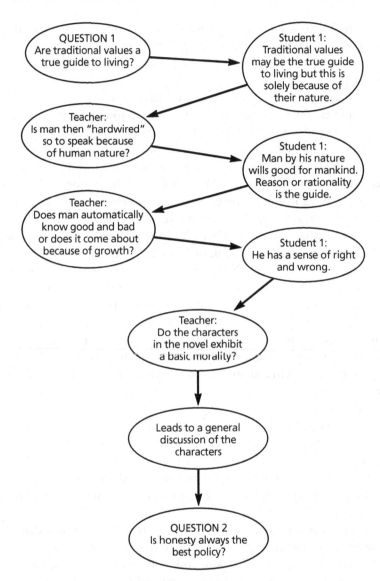

Figure 2.2 Discussion leading Question 1 and Question 2.

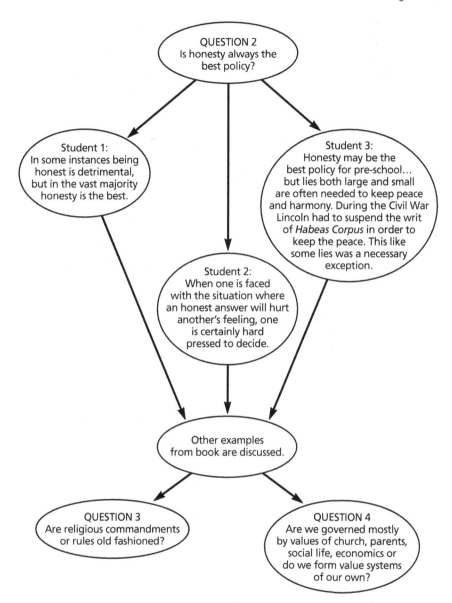

Figure 2.3 Discussion leading Question 2 to Question 3 and 4.

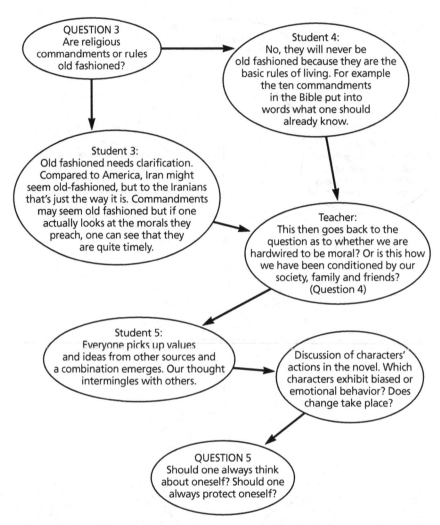

Figure 2.4 Discussion leading Question 4 and 4 to Question 5.

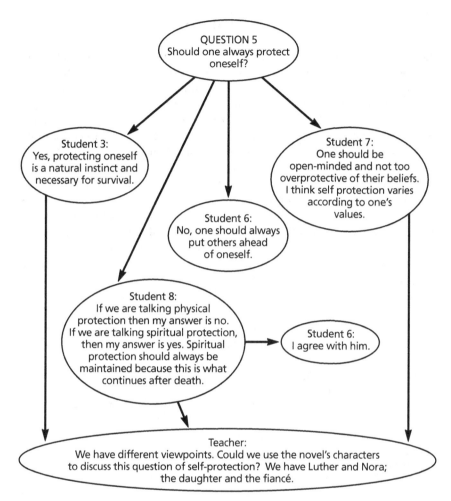

Figure 2.5   Discussion of Question 5.

CONCLUSIONS

The authors hope that they have conveyed to the reader the value of the use of conjecture-proof-refutation in the mathematics classroom and the use of critical thinking in the literature classroom in order to create meaningful learning experiences. The mathematics created by the students in trying to solve the classic 5-tuple diophantine problem clearly indicates that students are capable of original thought that goes beyond the mimicking and application of procedures taught in the classroom. Similarly, the critical thinking demonstrated by the students in the discussion of *Skipping Christmas* indicates that students are willing to discuss questions of belief, morality and values. The classic problem and the contemporary novel were tools used by the authors to sow the seed that allowed students to create mathematics and examine value systems. This method of teaching and learning adds personal meaning to the students schooling experiences, thus filling the existential void. As Bertrand Russell once said, "It should be one of the functions of a teacher to open vistas before his pupils, showing them the possibility of activities that will be as delightful as they are useful."

REFERENCES

Bacon, F. (1994). *Novum organum.* J. Gibson and P. Urbach (Eds.). Open Court Publishing Company.

Ernest, P. (1991). *The philosophy of mathematics education*, Briston, PA: The Falmer Press.

Fawcett, H. P. (1938) *The nature of proof.* New York: Teachers College, Columbia University.

Frankl, V. E. (2000). *Man's search for meaning.* Beacon Press.

Grisham, J. (2001). *Skipping Christmas.* New York: Doubleday Publishing Company.

Lakatos, I. (1976). *Proofs and refutations.* Cambridge, UK: Cambridge University Press.

Oystein, O. (1988). *Number theory and its history.* New York: Dover Publications.

Senk, S. (1985). How well do students write geometry proofs? *The Mathematics Teacher, 78,* 419–486.

Titus, H. (1994). *Living issues in philosophy.* Oxford: Oxford University Press.

Usiskin, Z. P. (1987). Resolving the continuing dilemmas in school geometry. In M. M. Lindquist, & A. P. Shulte (Eds.) *Learning and teaching geometry, K–12: 1987 Yearbook* (pp. 17–31). Reston, VA: National Council of Teachers of Mathematics.

APPENDIX A

Mathematics created by students via Conjecture-Proof-Refutation in the solution of the 5-tuple diophantine problem over a 7 month period.

Definition 1: A set of 3 positive integers $\{a_1, a_2, a_3\}$ is called a Diophantine 3-tuple if $a_i a_j + 1$ is a perfect square for all $1 \ i < j \leq 3$.

Conjecture 1: There exist infinitely many 3-tuples $\{a_1, a_2, a_3\}$.
    Proof: Let $a_1 = x - 1$; $a_2 = x + 1$ and $a_3 = 4x$
        Then $(x - 1)(x + 1) + 1 = x^2$
        Also, $4x(x - 1) + 1 = 4x^2 - 4x + 1 = (2x - 1)^2$, and
        $4x(x + 1) + 1 = 4x^2 + 4x + 1 = (2x + 1)^2$

Conjecture 2: There exist 4 positive integers $\{a_1, a_2, a_3, a_4\}$ called a Diophantine 4-tuple such that *at least one* $a_i a_j + 1$ is a perfect square for $1 \ i < j \leq 4$.
Laborious calculations yield: $\{1, 3, 8, 120\}$, $\{2, 4, 12, 420\}$, $\{3, 5, 16, 1008\}$

Conjecture 3: There exist infinitely many 4-tuples of the form $\{a_1, a_2, a_3, a_4\}$. The fourth integer a_4, must be some combination of a_1, a_2 and a_3 as found in Conjecture 1.
    Proof: Let $a_1 = x - 1$; $a_2 = x + 1$; $a_3 = 4x$; and $a_4 = 4x(2x - 1)(2x + 1)$.
        Then $a_i a_j + 1$ is a perfect square for $1 \ i < j \leq 3$.
        For $j = 4$, $a_1 a_4 + 1 = (x - 1) \ 4x(2x - 1)(2x + 1) = 4x(x - 1)$
        $(4x^2 - 1) + 1 = $ Hard to factor into a perfect square!

Students reach a dead end. This resulted in numerical calculations using the sets $\{1, 3, 8, 120\}$, and $\{2, 4, 12, 420\}$, in order to somehow factor $a_i a_4 + 1$ into a perfect square. One month later a new proof is attempted, based on student discovery that $120 = 1 + 3 + 8 + 2 (1)(3)(8) + 2 \sqrt{4} \sqrt{9} \sqrt{25}$, where $4 = (1)(3) + 1$; $9 = (1)(8) + 1$ and $25 = (3)(8) + 1$.

    New Proof: Let $\{a_1, a_2, a_3\}$ be a Diophantine triple and $a_1 a_2 + 1 = r^2$,
        $a_1 a_3 + 1 = s^2$, $a_2 a_3 + 1 = t^2$, where r, s, t are positive integers.
        Let $a_4 = a_1 + a_2 + a_3 + 2 \ a_1 a_2 a_3 + 2rst$.
        Then $\{a_1, a_2, a_3, a_4\}$ is a Diophantine quadruple, because:
        $a_1 a_4 + 1 = (a_1 t + rs)^2$,
        $a_2 a_4 + 1 = (a_2 s + rt)^2$,
        $a_3 a_4 + 1 = (a_3 r + st)^2$.

Numerous 4-tuples are verified and fit the pattern described in the new proof.

Conjecture 4: There exist 5 positive integers $\{a_1, a_2, a_3, a_4\ a_5\}$ called a Diophantine 5-tuple such that *at least one* $a_i a_j + 1$ is a perfect square for $1\ i < j \leq 5$.

Calculations last for over a month. Students use the sets $\{1, 3, 8, 120\}$, $\{2, 4, 12, 420\}$, $\{3, 5, 16, 1008\}$ but are unable to find a_5, such that all products + 1 yield a perfect square.

Conjecture 5: There does not exist a Diophantine 5-tuple $\{a_1, a_2, a_3, a_4\ a_5\}$, such that $a_i a_j + 1$ is a perfect square for $1\ i < j \leq 5$.

Students are unable to construct a general argument. This changes their conjecture to.

Conjecture 6: The set $\{1, 3, 8, 120\}$ cannot be extended into a Diophantine 5-tuple.

Proof: Suppose a_5 is the 5th number. Then the following equations must all be true at the same time:
$a_5 + 1 = r^2$; $3a_5 + 1 = s^2$; $8a_5 + 1 = t^2$; $120\ a_5 + 1 = u^2$, where r, s, t, and u are integers.

This means there exist perfect squares of the form:
$r^2 - 1$; $(s^2 - 1)/3$; $(t^2 - 1)/8$; and $(u^2 - 1)/120$.
Consider $r = 2, 3, 4, 5, 6, 7, 8, 9, 10, \ldots$
Then $r^2 - 1 = 3, 8, 15, 24, 35, 48, 63, 80, 99, \ldots$
It is impossible for $r^2 - 1$ to be a perfect square. This is also true for s, t, and u. Therefore a_5 does not exist.

Note: The proof created by the students for conjecture 6 does not resolve the 5-tuple problem by any means because one still has to check infinitely many other possibilities. However, the reader might appreciate the fact that 9th graders worked their way up to try and resolve a problem similar in difficulty to Fermat's Last Theorem.

APPENDIX B

The five anticipatory questions used in the discussion of the novel.

1. Do you think traditional values are the true guide to living?
2. Is honesty always the best policy?
3. Are religious commandments or rules old fashioned?
4. Are we governed mostly by values of church, parents, social life, economic conditions, political views etc., or do we form value systems of our own?
5. Should one always protect one-self?

CHAPTER 3

MATHEMATICS AND LITERATURE

Synonyms, Antonyms or the Perfect Amalgam?

Bharath Sriraman
The University of Montana

ABSTRACT

Mathematics is typically viewed as facts, procedures, and quantification and not associated with the activities of writing, discussing or reading literature. Literature on the other hand is used to expose young minds to various cultural and societal themes, historical ideas and to provide the context for critical thinking skills to develop and manifest. Mathematical and literature have one aspect in common, namely critical thinking, the process of making unbiased valid inferences. The preceding three sentences begs the question as to why mathematics and literature are regarded as antonyms when critical thinking is an attribute that is common to both fields. Can literature be used as a context for critical thinking as well as for introducing new mathematical ideas? The author explores this question in this paper by reflecting on the outcomes of using Flatland in a beginning Algebra course with 13–14 year-old students.

Interdisciplinarity, Creativity, and Learning, pages 31–39

31

INTRODUCTION

The 21st century student lives in an age where there is a plethora of readily available information about every conceivable or imaginable thing. For instance, the Internet is one of the gateways to a universe of information unimaginable just two decades ago. Even the technologically "unsavvy" cannot escape the barrage of information from the various media sources. It is reasonable to say that information saturation defines the human condition in the developed world. This leads to the question of our role as present day educators. The typical school curriculum is administered in discrete packages to the students, math, social studies, science, literature, languages etc. Yet reality does not function in this discrete manner, which leads to the following questions:

1. What are generic skills that cut across disciplines and are valuable in the real world?
2. Is there a pedagogical model available that can implement the cultivation of these generic skills?

Reflecting of these questions led to some insights. It is safe to say that problem solving and critical thinking are generic "thinking skills" that cut across a variety of disciplines. The mathematics education literature shows successful experiments by teacher educators at the university level using problem solving as a means to create meaningful learning experiences which students connect to personal metaphors such as "stories" or "games" and "adventures" (Chapman, 1997). Other approaches used by university educators involved creating problem-solving situations to enable students "to make isomorphisms between different problems to help them make convincing arguments to justify their ideas and promote critical thinking" (Glass, 2002, p. 75).

The development of critical thinking skills is traditionally one of the goals the liberal arts. Critical thinking is valuable not only to leaders who have to make difficult decisions but also used by businesspeople, historians, mathematicians, philosophers and scientists. Critical thinking has historically found expression in mathematics in the context of connecting ontological questions about the nature of mathematics to questions in theology. To expect pre-teenagers to contemplate like professional philosophers and theologians about ontological questions is unreasonable and "far-out"! Any well-intentioned teacher attempting to introduce critical thinking by synthesizing mathematics and theology can expect a bombardment of phone calls from irate parents as well as grief from administrative supervisors about the "questionable proceedings" in the classroom. How can critical thinking be implemented in a mathematics classroom with 13–14 year olds? This was the burning question confronting the author.

The Archimedean "eureka" moment manifested with the rediscovery of a precious book whilst organizing a cluttered bookshelf. The answer sought came in Flatland, a 19th century underground publication. The author of this book Edwin Abbott (1984) spins a satire about Victorian society in England by creating an isomorphic world called Flatland whose inhabitants are a hierarchy of geometric shapes and exhibit the many peculiarities of 19th century England.

The back cover of the book gives the following précis of the book.

> *Flatland* takes us on a mind-expanding journey into a different world to give us a new vision of our own. A. Square, the slightly befuddled narrator, is born into a place, which is limited to two dimensions—irrevocably flat—and peopled by a hierarchy of geometrical forms. In a Gulliver-like tour of his bizarre homeland, A. Square spins a fascinating tale of domestic drama and political turmoil, from sex among consenting triangles to the intentional subjugation of Flatland's females. He tells of visits to Lineland, the world of one dimension, and Pointland, the world of no dimension. (Back cover of *Flatland*)

The novel culminates tragically with A. Square committing heresy by preaching about higher dimensions such as the third and fourth dimensions. The personal choice of this book as "suitable" was re-enforced with the release of the book *The Annotated Flatland* (Abbot& Stewart, 2002) three months after the experiment was conducted. Renz (2002) in his review of the annotated book in the Scientific American wrote that the ideas from Flatland "anticipated ideas of today's string theory, one approach to fundamental particles. In addition to our three regular space dimensions plus time, this (string) theory postulates at least six more space dimensions curled so small as to be undetectable by current means!" (p. 90)

In hindsight this book seemed to be the ideal choice to implement the goal of cultivating critical thinking and introducing abstract mathematical ideas in a mathematics classroom. After working out the logistics with the school principal, a classroom set of Flatland was ordered for the Algebra course taught by the author, a full time mathematics teacher, at a rural mid-western public high school in the United States. One can imagine the initial reaction of the class with 23 students when copies of the book were handed out.

> "I thought this was Algebra class…why are we reading a book?"

> "This book is sexist…it talks about subjugating females"

The author set up this classroom experiment by explaining that the book was a work of fiction and would add some 'spice' to the math class. Since the class met everyday for 50 minutes, Thursdays were set aside for the book, and a portion of Friday was used to review, re-enforce, and expand the mathematical ideas connected to the reading. A reading schedule was provided to the class. The book was read over 4 weeks in November 2001. The fact that reading, writing and discussing the book replaced one of the tests in the class went over very well with the students. The remainder of the paper is devoted to presenting the classroom discussions generating from the book and to make the argument that mathematics fiction successfully "marries" literature and mathematics and creates the ideal forum to nurture critical thinking as well as introduce sophisticated mathematical ideas to 13–14 year old students.

Week 1: Social Distinctions, Polygons and Limits

The class was abuzz on the day the first discussion of the book was scheduled. Several students were explaining to others how to interpret the workings of Flatland. Several other students were engaged in a lively debate about the status of the working class and women in Flatland. These were good indications that the book was a good choice. Each discussion day began with students' "free" writing for 10 minutes about their thoughts on the reading and questions that were provoked by the book. The class was then split into six groups. Each groups' task was to reach a consensus on one question appropriate for discussion. These questions were used to moderate the classroom discussion. The questions this week related to understanding the hierarchy of shapes present in Flatland and the strange methods used by the inhabitants to distinguish one another. The discussion on various polygons was a thorough review of geometry covered in the earlier years. One student made the point:

> The lines in Flatland are not lines... they are actually very thin isosceles triangles. Imagine taking a big isosceles triangle and squeezing the base, making it narrower and narrower... it ends up looking like a line.

This comment created the perfect opportunity to introduce the analogous notion of a limiting process in mathematics. Since circles in Flatland were defined as regular n-gons with large values of "n," the author used this to introduce Archimedean constructions to create better and better approximations to the value of π.

On a non-mathematical note, several students remarked that the seemingly strange methods used by the inhabitants of Flatland paralleled the social structure of the school and society in general. This was a manifestation of critical thinking.

The way they "feel" each other to discover what class they are in is what we do now with appearances. If someone looks a certain way we will not associate with them . . . the emphasis is put on the appearance.

The fair abundance of 10° specimens, absolutely destitute of civic rights parallels our society treats poor people with no education with no regard and abuses their rights, like the treatment of the migrant workers.

Week 2: Comparing and Contrasting Flatland and Society today

The discussion from the previous week was fresh in the minds of the students and five out of the six group questions pertained to the living conditions of the inhabitants of Flatland, the structure of their society and the position of women in society. This class session was marked by a lively debate as summarized by the various contrasting viewpoints in the following classroom vignettes.

It (Flatland) is a lot like today because the lower classes get a poorer education, little respect, and do most of the work. The higher classes get a good education, therefore they can use this to gain respect and higher standings. This is true of our society today because the lower classes also go to poorer public schools where they don't learn as much because of lack of facilities . . . so they don't go as far in life and can never make it to the higher classes. The higher class continues to climb and gain more knowledge, power and money by getting a better education at private expensive schools.

The upper class in both worlds think they are the best. In today's society . . . on paper the rich do not have more rights but it seems like they do. If the upper class break laws they have to face the same consequences . . . although they may be able to afford better lawyers and get away with some loophole of the other.

There are different classes in our society but they are not set apart by rules, they are equal and have the same opportunities and freedoms.

In today's society we do not consider our priests the highest class although it used to be that way a long time ago. Also women and men are considered equal by our society, it is not so in Flatland.

Fashion as in the "swaying of the back" in public is parallel to how fashion plays out in society now. People have to "look" a certain way in public now. Even today in our society woman are looked at as the weaker gender.

Flatland is a very odd place to live. The people aren't racist . . . they are "shapist," discriminating on the basis of shape. As a girl I would feel discriminated against, having to keep moving constantly according to the laws of Flatland.

Our school is a lot like Flatland because many girls think they need to dress or act a certain way in order to get accepted. So our society may not be racist but it sure is "shapist."

Week 3: Dimensions and Dilemmas

The third week was characterized by a discussion of the notion of dimension. Just as A. Square, the protagonist from Flatland somehow realized that higher dimensions existed (such as our world: Spaceland), students raised the possibility of the existence of higher dimensions. Students were divided about the existence of a fourth dimension, until one of the groups mentioned that time could be considered as the fourth dimension. Another student remarked that if one thought of dimension as a mathematical concept, we could have as many dimensions as we wanted.

This discussion was the perfect stimulus to present simple notions from relativity and to introduce the model of Minkowskian space -time geometry. The students were fascinated with the idea of objects not having length in an absolute sense and length being dependent on the observers' frame of reference.

The class pursued the analogies used in Flatland to illustrate higher dimensional objects existed. One group argued for the existence of 4-dimensional objects as follows:

A point has no dimension. Take two points and connect with a line, that is 1-dimensionsal. Take two lines, connect them to get a square, which is 2-dimensional. Two squares give a cube, which is 3-D. Now two cubes connected should give the next level which is 4-D.

Another group made the following remark about the construction outlined in Figure 3.1.

The geometric sequence that results from the construction (Figure 3.1) 1 point (dot), 2 points (line) 4 points (square), 8 points (cube), 16 points (hyper-cube) . . . shows that the construction can be carried on as long as you want and shows its mathematically possible to have very high dimensions.

As indicated in Figure 3.1. Flatland brought forth a rich discussion on the idea of dimension and led the students to imagine new possibilities. The author also introduced the idea of decomposing objects into a certain number of self-similar pieces to produce "fractal" dimensions from fractal geometry. The analogy of moving "up" dimensions by using self-similar ob-

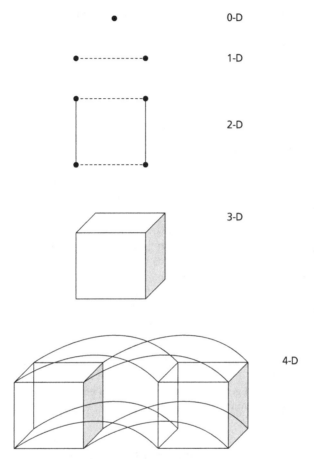

Figure 3.1 Construction of a 4-dimensional object by analogy.

jects made the students open to moving "between" dimensions when constructing fractals. One student remarked

> The Sierpinski triangle is a figure between a line and a triangle because it is all broken up, so it makes sense that it has a dimension between 1 and 2.

This is a very sophisticated mathematical idea being expressed by this student. Some other students tried to draw a distinction between "real" dimensions versus "mathematical" dimensions by trying to connect the notion of dimension to the degree of polynomials. Some students who were fans of the TV series *StarTrek* were very imaginative and remarked.

It's really easy to convince a spacelander (us) that a 4-D world exists. We simply have to travel through a worm hole and on the other side will be a hyper-cliff consisting of little hyper-spheres, hyper-cubes, hyper-triangles and the spacelander would then believe.

I believe one (the fourth dimension) would exist but it would be like a person within a person.

Week 4: The Next Frontier

The final discussion revolved around the main themes of the book. Many of the students were disappointed by the tragic ending of the book and had some startling insights to analogies in society. The questions revolved around the notion of free speech and expression and instances where people were persecuted because of their beliefs or ideas. The discussion was "heavy" especially when some students pointed towards examples from history that were analogous to the tragic end of A. Square.

A. Square was ahead of his time. It is a lot like the person[1] who said that the earth is not the center of the universe and the church imprisoned him.

There was this movie[2] where they threw this guy in prison because he was a genius and knew every little thing about the government and all their secrets.

...Like Stalin started getting rid of all these intellectuals because they were writing against communism.

CONCLUSIONS AND IMPLICATIONS

The process of critical thinking could be viewed as the use of reasoning in the pursuit of "truth." Critical thinking has also been defined as the ability and tendency to gather, evaluate, and use information effectively (Beyer, 1985). The students in this Algebra course demonstrated their capacity for critical thinking by making inferences on "truths" about society and life as is evidenced in the various classroom vignettes. The book became a didactic tool of sorts and provided the ideal scaffolding for students to critically examine the norms and biases of society. The students were reasoning by analogy, as well as using their imagination. The ideas in Flatland also exposed students to some very advanced mathematical ideas such as dimension. This created the perfect setting for the teacher (author) to expose students to broader mathematical notions such as limits, non-Euclidean geometries such as the Minkowskian space-time geometry and Fractal geometry, and historical techniques of approximation. Although the students were discussing the novel, there was a general inquisitiveness about a variety of societal issues and new mathematical ideas. Students were willing to listen to alternative opinions and most importantly were willing to examine their own biases and prejudices (Facione, 1990). The experiment with Flatland was talked about for the rest of the school year and some of the students suggested in reading one of the many sequels to the book. This was done in the second semester but space constraints prevent another exposition at this juncture. Perhaps a sequel will follow!

REFERENCES

Abbot, E. (1984). *Flatland* (Reprint of the 1884 edition). Signet Classic Books.

Abbott, E., & Stewart, I (2002). *The Annotated Flatland.* Perseus Publishing.

Beyer, B.K. (1985). Critical thinking: What is it? *Social Education.* 49, 270–276.

Chapman, O. (1997). Metaphors in the teaching of mathematical problem-solving. *Educational Studies in Mathematics, 32,* 201–228.

Facione, P. (1990). *Critical thinking: A statement of expert consensus for purposes of educational assessment and instruction.* The California Academic Press, Millbrae, CA.

Glass, B. (2002). Students connecting mathematical ideas: possibilities in a liberal arts classroom. *Journal of Mathematical Behavior, 21,* 75–85.

Renz, P (2002). *Scientific American. 286*(4), 89–90

NOTES

1. The student was referring to Copernicus.
2. There is a whole genre of movies where highly intelligent people who stumble upon classified information are imprisoned under the pretext of "national security."

ACKNOWLEDGMENT

Reprint of Sriraman, B. (2003). Mathematics and Literature: Synonyms, Antonyms or the Perfect Amalgam? *The Australian Mathematics Teacher, 59*(4), 26–31. Reprinted with permission from the Australian Association of Mathematics Teachers © Bharath Sriraman.

CHAPTER 4

MATHEMATICS AND LITERATURE (THE SEQUEL)

Imagination as a Pathway to Advanced Mathematical Ideas and Philosophy

Bharath Sriraman
The University of Montana

INTRODUCTION

This article is the sequel to the use of *Flatland* with beginning algebra students reported in Sriraman (a, in press). The use of *Flatland* with beginning algebra students resulted in the positive outcomes of cultivating critical thinking in the students as well as providing the teacher with the context necessary to introduce sophisticated mathematical ideas. The marriage of mathematics and literature led students to reflect on contemporary society and its problems as well as gain an insight into notions of limits, historical approximation techniques and various non-Euclidean geometries (Fractal geometry and Minkowskian space-time geometry). This atypical but re-

Interdisciplinarity, Creativity, and Learning, pages 41–51
Copyright © 2009 by Information Age Publishing
41

freshing learning experience led students to request the reading of one of the sequels to *Flatland.* The "providential" release of Stewart's *Flatterland* in 2001 seemed like the ideal follow up to Flatland. Banchoff's (2001) review of *Flatterland* for the Mathematical Association of America partially found in the back cover states: "*Flatterland* challenges readers to go beyond *Flatland* and deal with phenomena, not just in dimensions higher than four, but in many exotic geometric realms that stretch our imagination and powers of visualization." Upon reading the book over the winter break, my personal impression was that the ideas introduced in the entire book would pose a challenge to university math and physics majors. However the material in the first five of the eighteen chapters were within the scope of 13–14 year old ninth graders. In fact some of the ideas introduced in the first five chapters such as arbitrary dimensions in mathematics, and fractal geometry had been discussed in the class during the reading of *Flatland.* In addition Stewart (2001) had brilliantly made modern ideas such as encryption on the Internet, the taxi-cab metric, and fractal dimensions, among others very accessible to the lay person. This was achieved by creating a contemporary setting in which the heroine Vikki, the great-great-grand-daughter of Flatland's protagonist A. Square, is taken on a guided virtual reality tour of the mathematical universe by a space hopper.

SETTING UP THE EXPERIMENT

I was fortunate once again to have the support of the principal in this endeavor and was supplied with a classroom set of *Flatterland.* I decided to make use of the book towards the end of the high school year, when interest in the regular curriculum typically begins to wane. The pedagogical reason for doing this was to ensure that students had the mathematical background (from the algebra curriculum) necessary to allow me to develop the ideas in the book. My goal was to use *Flatterland* as scaffolding in order to:

1. explore non-intuitive math problems,
2. further students' understanding of dimension,
3. to help students gain a deeper understanding of fractal geometry, and
4. to develop the taxi-cab geometry.

The first five chapters of *Flatterland* were read sequentially in April-May 2002. The reading structure was very similar to that used in the reading of Flatland. As in the experiment with *Flatland* (Sriraman, a in press) , reading, writing and discussing the book replaced one of the tests in the second semester, which again went over very well with the students.

I explained that the book was a recently released fun sequel to *Flatland.* The five chapters were read over eight weeks. A reading schedule was provided to the students. We discussed and developed the ideas from the first five chapters in eight 50-minute Friday class periods. Unlike the previous experiment where students were split into groups for discussion, I used a show of hands to separate the class into two groups, namely those that "thought" they understood the reading and those that were confused. Then I applied the Socratic method of *question-hypothesis-elenchus -acceptance/rejection* to moderate the classroom discussion. In other words classroom discussion began with a "confused" student stating the nature of their confusion, which was then restated as *question.* Then non-confused students were asked to respond to the question. Their response/explanation was used to generate a testable *hypothesis,* which was subject to *elenchus* (or refutation), eventually leading to *acceptance* of the explanation or *rejection,* in which case the hypothesis was re-examined. This process was modeled many times in the regular algebra curriculum in order to set the stage for the discussion of *Flatterland.* The fascinating outcomes of the use of *Flatterland* with the 13–14 year old ninth graders is presented in the next sections.

"TIMES ARE A CHANGING"[1]

The first three discussions revolved around the contemporary setting of *Flatterland,* the dashing heroine Vikki, and strange packing problems. Having read *Flatland* in the first semester, students were able to use to the contemporary setting of the book and compare the changes described in the book to the tidal waves of change in the 20th century. First, the girls in the class were pleased that the protagonist was a female, who was roughly their age. Second, the students were happy to see that the chauvinistic society of *Flatland* had evolved and embraced women's liberation due to various Flatlandian wars and revolutions. Third, students really enjoyed the satirical word play of the book and relished the double-edged nature of words used by Stewart (2001). Student comments that made various critical comparisons on the changes in Flatland and their parallels to our world are summarized below.

> (In Flatterland) Flatland has changed in quite a few ways. They have bcome more intelligent and not so prejudiced against women. Women are treated as equals rather than just unintelligent things. They send messages like us (e-mail) and code stuff...I'd say Flatterland is the more mature version of Flatland.

> Flatterland now has interline computers like our internet computers, telephones and many other devices that were actually invented in the past cen-

tury. Women have gained more respect and rights in Flatterland just like the voting movements (suffrage) in our history...but they still have to sway their back from side to side and sing to avoid hurting other people.

It still seems as though people won't accept the fact of there being a 3rd dimension. You are deemed "weird" if you have a radical idea, and it still happens today in our society when people don't accept new ideas from science.

Did you notice that generations now do not gain another edge with each generation. I like the word play, like you don't have an edge over others just cause you're born into a rich family. But we all know in reality you do have an edge if your parents are rich.

IS EVERYTHING MATHEMATICALLY POSSIBLE? PACKING FRUIT AND DEMENTIA WITH DIMENSIONS

The third chapter of *Flatterland* describes the visitation of the "Space hopper" and Vicki's discovery of the space hopper's strange shape. The use of a time-series of cross-sections was understood by the students and thought similar to the discovery of the Circular visitor's shape by A. Square. Just as a sphere moving through a horizontal plane would first be seen as a dot followed by a series of expanding circles and then by a series of contracting circles culminating in a dot and then disappearance; students were able to extend this notion and accept that it was plausible to think of a series of expanding and contracting moving spheres as the shadow a 4-dimensional hypersphere would cast in our world. Students found some of the ideas in the third and fourth chapters a little difficult to understand. For instance Stewart (2001) describes the non-intuitive possibility of fitting a cube of side length 1.06 into a unit cube! One of the students actually tried to accomplish this by using thin cardboard boxes but was unsuccessful. This was discussed in the class and students concluded that it was practically "impossible" to find instruments that could be used to make a cube of side length 1.06. Therefore they deemed this non-intuitive notion of squeezing a larger cube into a smaller one as "mathematically" possible but practically "impossible." I used this opportunity to discuss the need for calibration and accurate measuring instruments in science to generate data that could verify or refute hypothesis. Many students were also unable to completely understand the encryption procedure described by Stewart (2001) to encode and decode messages. However this presented the opportunity to introduce the binary numeral system and play with the four basic arithmetic operations in base 2. Students really enjoyed this and realized the arbitrariness of numerals.

Students were also intrigued and understood the fruit packing problem described by Stewart (2001), namely what is the most efficient way to pack fruit (that are roughly spheres) into cardboard boxes that are rectangular prisms? This problem allowed us to explore approximation techniques to determine the maximum amount of a fruit that could be packed into a box, given particular starting assumptions of the size of fruit, and the size of boxes. We choose starting sizes for apples, grapefruit, pumpkins and watermelons to determine how many could be packed in a standard fruit box found in the grocery store. One student took the initiative of going to a grocery story and bringing some perforated sheets used in apple boxes, which visually demonstrated the practical nature of the packing problem. Student comments follow:

> It seems that it is mathematically possible to do anything. I don't believe you can fit a bigger cube into a smaller one... But I really like the idea of using binary digits for changing letters to symbols with only 0 and 1 and it made sense how to spot the errors.

> In Spaceland or "Planeturthian" you have more space than in Flatland. They are forced to stick oranges with gaps in Flatland... but we can stack them differently and reduce the gaps because we can move in more directions than they can.

> It makes sense to add a dimension every time you can move something in a different direction. So we can have a "chalk-cheese" direction just like north-south, east-west, up-down. The idea of finding the dimension also makes sense because if you had a 3D ball, its surface is 2D, because 3–1 =2. So the surface is always one dimension smaller than the original dimension. So a 101 dimensional ball has a 100D surface.

> The idea of stacking spheres to make a hypersphere is really far out. But it makes sense if you think of making a sphere by stacking together smaller and bigger circles

FRACTALS, TAXICABS, AND SQUARE CIRCLES

The fourth chapter of *Flatterland* describes the arbitrary nature of dimensions in mathematics by surveying ideas from different geometries. In *Flatland,* the notion of self-similarity was used by the circular visitor from Spaceland to postulate the existence of higher dimensions to A. Square. This notion was understood by students as was evidenced in their constructions of 4-dimensional objects such as a hyper-cube and in their argument that the geometric sequence 1,2,4,8,16... that counted the number of vertices of self-similar objects indicated the existence of higher dimensional objects (even if one could not visualize them). During the previous experiment with *Flatland*, I had used students' notion of self-similarity to construct the Sierpinski triangle and posed the question about the dimension of fractal objects. In *Flatterland*, the notion of fractal dimension is explored at a deeper level. One of the classroom discussions revolved around the calculation of fractal dimensions. Table 4.1 summarizes student strategy to calculate fractal dimensions.

Student's calculated the dimensions of the Sierpinski triangle and the Koch snowflake using trial and error on their calculators. Since they were not exposed to the notion of logarithms, I decided that it was pedagogically sound to perform this calculation by trial and error. The dimensions found for the Sierpinski triangle and Koch snowflake were 1.584 and 1.261 respectively, which was accepted by the class as being accurate. One industrious student located a fractal web-site, which we used to check our dimension calculations.

In the consequent discussion of the Mandelbrot set, Stewart (2001) makes use of the taxicab metaphor to describe co-ordinates in the complex plane. The description given to Vikki about the moves necessary to get from

TABLE 4.1   Extension of Self-Similarity Idea to Discovery of Formula for Calculating Fractal Dimensions

Object	Dimension of Object	Self-similar copies made	Pattern
Point	0	1	$1 = 2^0$
Line segment [2 points]	1	2	$2 = 2^1$
Square [two segments] [4 points]	2	4	$4 = 2^2$
Cube [Two squares] [8 points]	3	8	$8 = 2^3$
Hypercube [Two cubes] [16 points]	4	16	$16 = 2^4$

[*Student observation: The dimension always shows up in the exponent*]

Sierpinski Triangle	don't know	3	$3 = 2^{dimension}$?
Sierpinski Triangle	Know (1.584)	3	$3 = 2^{1.584}$
Koch Snowflake	don't know	4	$4 = 2^{dimension}$?

[*Student observation: The dimension of the Koch Snowflake can't be the same as a square. The size of the snowflake doesn't double, but triples every time*]

Koch Snowflake	Know (1.261)	4	$4 = 3^{dimension}$ $4 = 3^{1.261}$

one point to another in *Quadratic City* created the perfect setting to talk about Taxicab geometry. Since students had been exposed to the distance formula in the analytic geometry segment of the algebra curriculum, it was quite easy to introduce the taxi-cab metric by analogy.

We explored several questions related to *Flatland* and middle school geometry, namely are there any regular polygons in Taxicab geometry? In particular we compared the notions of betweenness in the Euclidean and Taxicab metrics and explored various Flatland shapes such as isosceles triangles, squares, and circles in the Taxicab metric. A vignette of this exploration follows.

A Classroom Vignette

Teacher: How do we determine whether one point is between two other points?

Student 1: You plot the points and see where they are on the real number line.

Student 2: Couldn't we use the mid-point formula?

Student1: But you can't be sure that this point in exactly in the middle of the two other points.

Teacher: Don't we need some information about the location of the points to use the midpoint formula?

Student 3: How can you be sure that the points are on the real line? Can't the points be outside the real line?

Student 2: Yeah, we use (x,y) co-ordinates to locate the points.

Teacher: Why don't we look at an example. What if we take the points P(3,2), Q(6,4) and R(9,6) and plot them.

Student 4: The points are on the line with slope 2/3.

Teacher: Okay, now how do we check whether Q is between P and R?

Student 1: Just use the plot and you see the point Q is right in between.

Teacher: What if you don't have graph paper and you can't plot the points?

Student 2: You can just visualize it in your head.

Teacher: Can we use any formulas we learned ?

Student 4: The mid-point formula?

Teacher: But you can't always be sure that one point will always be between the other two.

Student 5: Why don't we look at distances between the three points?

Teacher: That's a good idea. Does anybody remember how we calculate distances?

Student 6: The distance formula.

Teacher: Yes, but how can we use the distance formula to decide that Q is between P and R?

Student 6: Calculate PQ, then QR, and then see if they add up to PR.

Teacher: Does anybody disagree with this idea?

Student 4: Yeah, but does it always work?

Student 6: I think it does. You can check it on the real line if you like.

Teacher: If it works on the real line, do you think it works on every line?

Student 6: Yeah, cause the real line is just another line with no slope.

[Students perform calculations and determine that PQ + QR = PR]

Teacher: Can we define betweenness now?

Student 6: We already did. Just calculate the three distances and see if the two smaller ones add up to the total distance.

Teacher: Okay, so we say that a point B is between A and C if AB + BC = AC, and to make life easier we write A-B-C. Now the question is does it work the same way in the taxicab world?

Student 5: But how are we going to calculate distances there? Don't we need that?

Student 7: You can just count on a graph paper.

Teacher: Can we come up with a formula maybe? Just like the distance formula we already know? $d = \sqrt{(x_1 - x_2)^2 + (y_1 - y_2)^2}$

[Classroom discussion eventually leads to the discovery that distance in the taxicab world is calculated by counting the number of blocks traveled either east-west plus the number of blocks traveled up-down]

Teacher: We can write this formula as $d_T = |(x_1 - x_2)| + |(y_1 - y_2)|$ and we'll use d_E for the normal way of calculating distance. Now can we check whether Q is between P and R?

Student 8: Do we do the same thing like before?

Teacher: What do you mean?

Student 8: Like see if those three distances add up?

Teacher: What does the class think?

Student 6: I think it will be the same.

Teacher: Same what?

Student 6: Like the same rule, you know AB+ BC has to equal AC.

[Calculations reveal that betweenness does work out the same way]

Student 7: Can't we go from P to R like in a city, where you are trying to avoid a block. What I'm saying is can't you go through a different point, like some point X and get to R. Does that mean there are other points P and R that are between but not on the line?

[*This elenchus (refutation) led to a discussion of the difference between "metric" betweenness and betweenness as defined in Euclidean Geometry. The discussion led us to re-examine the Euclidean hypothesis for betweenness and reach the following conclusion*]

Teacher: We will impose the requirement in the definition of betweennness that points be on a particular line to take care of this problem of "metric" betweenness in Taxicab geometry. This way we can use the same definition in both geometries

Commentary on Vignette

This vignette is used for the purpose of illustrating the Socratic method. As is evident in the transcribed comments, students were unwilling to accept statements made by other students and the teacher blindly but subjected it to scrutiny. As teachers we should value the pedagogical value of the Socratic method even though it can be very tedious on occasions. The preceding vignette is a condensed and edited version of 45 minutes of dialogue eventually leading to the acceptance of the Taxicab metric and the notion of betweenness. In the discussion in the following week we used the taxicab metric to explore regular polygons and circles.

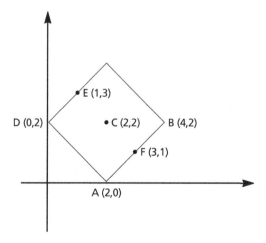

Note that CD = CE = CB = CF = CA = 2

Figure 4.1 A "square" circle in Taxicab geometry.

Students were "blown away" by the bizarre appearances of known Euclidean objects in the taxicab metric. In addition to discovering circles appeared as squares in the taxicab metric, the class also found several known triangles to have strange properties. For instance equilateral triangles in the Euclidean metric turned out to be isosceles but equiangular in the taxicab metric!!

A FORAY INTO PHILOSOPHY

Towards the end of the fifth chapter, one of the students in the class asked the ponderable question "Is mathematics real?" The fact that Vikki was zipping in and out of different geometrical worlds in mathiverse (the mathematical universe) led many students to wonder whether mathematics was invented or discovered. In other words did Vikki discover geometries that were present a priori or were the different geometries a figment of the space hopper's imagination made real via the use of virtual reality. These questions were raised by several students who were big fans of the movie *The Matrix* in which reality as is was quite different from reality experienced through a virtual interface. These questions posed the challenge of introducing mathematical philosophy at an elementary level. The following vignettes illustrate 13–14 year old student' viewpoints on the nature of mathematics. I have condensed two students' expressions of their viewpoints, which parallel the Platonist and Formalist viewpoints of the nature of mathematics. It is interesting to note that the class was evenly divided between the Platonist and Formalist camps as in professional mathematics, which I believe was a function of the non-intuitive mathematics brought alive by the book.

> Mathematics is something real. I strongly feel this way because you can never prove math wrong. Every equation in mathematics has some connection to the real world. Let's use area of a square or a rectangle for example, you can always use this in the real world...and I'm sure that there is a mathematical equation that solves the packing orange problem. So for the various reasons I have stated I believe the mathematical world to be real. [This quote shows similarities to the Platonist view of mathematics]

> I believe mathematics is imagined. For one nothing on earth is perfectly predictable...like take the weather. I don't think there is any mathematical equation that can predict the weather. Mathematics is simply there to entertain people...it sometime helps explain things like how to calculate area without measuring, and there are laws like Newton's laws of motion that solve some problems but not all. Math is food for thought...it can interest some people and easily scare others away. [This quote shows similarities to the Formalist view of mathematics]

CONCLUSION

The experiment with *Flatterland* lasted eight weeks towards the end of the school year. Just when student interest in the regular algebra curriculum had begun to wane, the book served as a catalyst to renew their enthusiasm for mathematics. The anticipation of being able to talk about the book on Friday kept student interest in the regular algebra curriculum alive. The ideas in the book deepened many students' interest in mathematics. It led three students in deciding to attend a summer math camp at a nearby university. Four of the students in the class are strongly considering a career that is related to mathematics or computer science. It is professionally satisfying to realize that the use of mathematics literature with students can led to vistas unexplored and unimagined both for the students and the teacher.

REFERENCES

Abbot, E. (1984). *Flatland* (Reprint of the 1884 edition). Signet Classic Books.
Banchoff, T. (2001). Read this! Review of Flatterland in *The MAA Online Book Review Column*. www.maa.org/reviews/flatterland.html
Sriraman, B. (2003). Mathematics and Literature: Synonyms, antonyms or the perfect amalgam. *The Australian Math Teacher*. [see previous article]
Stewart, I. (2001). *Flatterland*. Perseus Publishing.

NOTE

1. The title of a revolutionary Bob Dylan song released during the Civil Rights movement in the U.S.

ACKNOWLEDGMENT

Reprint of Sriraman, B. (2004). Mathematics and Literature (the sequel): Imagination as a pathway to Advanced Mathematical Ideas and Philosophy. *The Australian Mathematics Teacher, 60*(1), 17–23. Reprinted with permission from the Australian Association of Mathematics Teachers © Bharath Sriraman.

SECTION II

MATHEMATICS AND PARADOXES

CHAPTER 5

1 OR 0?

Cantorian Conundrums in the Contemporary Classroom

Bharath Sriraman
The University of Montana

Libby Knott
The University of Montana

1. INTRODUCTION

In set theory, one comes across the notion of "vacuous truth." A statement is vacuously true if it is true but doesn't quite say anything. The structure of a vacuously true statement is typically of the form: Everything with property A also has property B, with the caveat being that there is *nothing* in property A. For instance we could say: All humans with gills are sharks. This statement is vacuously true because there are no humans with gills. It is natural to dismiss such examples as absurd and pathologies within the framework of set theory. However the notion of vacuous truth arises in some pedagogical situations. The reader is undoubtedly curious whether a situation requiring the examination of "vacuous" truth can arise in a contemporary math-

Interdisciplinarity, Creativity, and Learning, pages 55–60
55

ematics classroom. In fact such situations do arise! One such situation in a pre-service elementary math classroom will be described, discussed and analyzed for classroom implications. Cantor's seminal work, which showed infinity comes in various sizes (!) was instrumental in the development of modern set theory. Cantor's contributions can be viewed as the Ur[1]-source of the consequent paradoxes that arose within set theory. In his honor, we will label unusual set-theoretic pedagogical situations *Cantorian Conundrums.*

2. THE SITUATION

In the United States, prospective elementary school teachers are required to take some math content courses at the university level which typically cover some elementary set theory to act as a foundational base and a context out of which models for the four arithmetic operations $(+, -, \times, \div)$ are developed. The situation we are about to describe took place in one such course. The following problem was part of the homework assigned to the students. At this stage of the course students were familiar with set theoretic notions of subsets, unions, intersections, complements etc.

> *Suppose B is a proper subset of C. If n(C)* ▪ *8, what is the maximum number of elements in B? What is the least possible number of elements in B?* (Billstein, Libeskind & Lott, 2004, p. 74).

At first glance the problem appears as a rudimentary set theory problem. However the ensuing discussion of this problem in the classroom revealed otherwise. The entire class agreed that the answer to the first part of the question was 7. In order to justify this, students appealed to the definition of a proper subset and discarding the one subset, namely the set itself. So far so good! When we got to the second part of the problem, a show of hands revealed that the class was divided 12–9 about the answer. Twelve students thought the answer was zero, whereas 9 students claimed that the answer was one. This was an interesting turn of events and the ideal moment to step back and ask students in either camp to justify their claims. The identity of the students in the respective camps was also noted. The more vocal of the students in the "Zero camp" essentially argued that one simply lists out all the proper subsets of the given set, and observes that the empty set is one of these proper subsets. Since there are no elements in the empty set, zero is the least number of elements that B can have. On the other hand, students in the "One-camp" argued "If the answer is zero, then aren't we viewing the empty set as an element? We cannot view the empty set as an element, so the answer should be one" (Class comment). Towards the end of the class period, the class was still far from reaching a consensus. This was the ideal opportunity to let students in either camp reflect on their choice. The "positivist" turn of events was too good a pedagogic opportunity to pass up. Hence following written assignment was announced.

3. TAKING AND DEFENDING A POSITION

Choose a position (1 or 0). Based on your position, write a two-paragraph "opinion" defending the validity of your choice.

The written opinions, collected three days later, revealed a 14–6–1 split. Fourteen now favored 0, six of the nine original students still favored 1 (no switches), and one student from the One-camp had shifted to "an answer dependent on vacuous or logical truth" favoring zero. Some sample responses from the three camps are now presented. These will help illuminate the subsequent discussion and implications sections of the paper.

4. THE "ONE CAMP"

"My position supports the "1 side"... the question is: is the empty set an element? No it is not. The definition (of proper subsets) states: "every element of B is contained in A and there is at least one element of A that is not in B." However this definition does not state anything about the empty set... the question asks for the number of elements, an empty set is not an element. Therefore the answer is 1"

"The empty set is a set with no elements. The empty set is not an element in a set but a subset or a set in general... Since the empty set is not an element but a set with no elements it cannot be considered (for) the least number of elements"

"I am guessing... that the answer is 1. I would have to say this because you have to have one element from C to be a proper subset at all. See figure:"

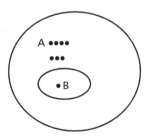

"Emptiness cannot be viewed as a common property between sets."

"The definition of an element is: individual objects in a set. Although the empty set is contained in each set, it is not one of the objects we take into account when viewing the number of objects. Since B is a subset of C, it has to always have at least one object in common with the original set C. The following example conveys this idea: Set C = {a,b,c,d,e,f}, Set B = {a}. In both these sets the empty set does not appear directly in the set."

"Suppose we start with the empty set ∅. How many elements does a subset of this set have? It is the set itself! So the answer is 1."

5. THE "ZERO CAMP"

"The set in question has a cardinal number 8, meaning there are 8 elements in set C. Let's say C = {1,2,3,4,5,6,7,8}, and we are going to list all the subsets of set C. The subsets of C are { }, {1}, {2}, . . . I am not going to keep going because the question is already answered. The empty set is a subset of C and { } contains zero elements. Therefore zero is the *correct* answer"

"I believe the answer to this question can be best described in a story . . . Pretend you have a bag of marbles. The number of marbles in this instant doesn't matter, what matters is that inside the bag . . . is a group of elements. These can be grouped even if you make a group of 0 elements. So as told in this marble story, you can have 0 elements because everything that is grouped inside the bag is still considered an element."

"Here is an example to prove that zero is the correct answer. If a set contains only one element, f, then its subsets include the empty set, which has zero elements, and {f}, which is only one element. The only proper subset in this situation is the empty set. Therefore zero is the least number of elements."

6. THE CAMP BETWEEN CAMPS

"The sum of the empty set is 0 (the nullary sum). Therefore the cardinality of the empty set is 0. For any set, in this case C, the empty set is a subset of C. Hence B, as a proper subset of C, could be the empty set. For example, if a set is a bag of elements, an empty bag may be empty (0 elements) but it still exists! There also exists a loophole called vacuous truth: a truth while "true" doesn't say anything. This stated, then, a set is finite if its cardinality is a natural number. But 0 is not a natural number. Therefore our vacuous truth is: the cardinality of the empty set is 0, the empty set is finite . . . So while the empty set itself contains no elements to speak of, in and of itself, it still exists as a subset. So logically speaking the answer is 1, and vacuously speaking the answer is 0."

7. DISCUSSION AND IMPLICATIONS

As the written responses reveal, the problem was a source of considerable contention, introspection, and led students to a careful re-examination of the definitions under use. If the goal of teacher training is to produce teachers that memorize and regurgitate mathematics, such an exercise could be construed as a waste of time. For those that value the final "positivist" black and white product, the answer is the end-all. On the other hand if the goal of teacher training is to create teachers who value the role of argumentation in the making of mathematics, i.e., the process over the product, such an exercise is invaluable towards pursuing this pedagogic goal. Although there were flaws in the reasoning of the "One camp," such as considering the empty set as an element of every set as opposed to a subset, many of these students stumbled upon *Cantorian conundrums* in this process. Questions such as "Can emptiness be viewed as a common property between sets" and "How many elements does the subset of an empty set have? are more philosophical in nature, than they are mathematical! The underlying affective hope here is that prospective teachers' experiences of an university educator valuing their thoughts will translate into them as teachers valuing young children's thoughts. The philosophical questions posed by the students are also an avenue through which the university educator can initiate an exploration of related set theoretic topics via projects. For instance the question: What are the subsets of the empty set, leads into a better grasp of the notion of power sets. Since every set has the empty set as a subset, the empty set is one of those very sets. Therefore it has itself as a subset. Which means having itself as a subset leads to itself as a member of its power set! In conclusion, we are reminded of Mason's (2003) recent exposition in AMT on the structure of attention from a cognitive standpoint. Although the context here is more argumentative, Mason's framework is certainly applicable from the observer's standpoint. In students' responses, one is offered an insight into "the lived experience of mathematical thinking," and observes to an extent the structure of their attention on abstract notions such as, elements of a set, subsets and the empty set. One conjectures that there is a change in the structure of attention about elements of a subset as evidenced in the number of students moving from the One-camp to the Zero camp! As an added bonus, the student in the "camp between camps" truly experienced the legacy of *Cantorian Conundrums!*

REFERENCES

Billstein, R., Libeskind, S., & Lott, J (2004). *A Problem Solving Approach to Mathematics for Elementary School Teachers (8th Edition)*. Addison Wesley Longman.

Mason, J. (2003).On the structure of attention in the learning of mathematics. *The Australian Math Teacher, 59*(4), 17–25.

DEDICATION

To our mutual understanding of the *Principle of toxicity.*

NOTE

1. In German and Swedish the syllable and preposition *ur* means "something original" or "the source"

ACKNOWLEDGMENT

Reprint of Sriraman, B. & Knott, L. (2006). 1 or 0? Cantorian conundrums in the contemporary classroom. *Australian Senior Mathematics Journal, 20*(2), 57–61. Reprinted with permission from the Australian Association of Mathematics Teachers, © Bharath Sriraman

CHAPTER 6

UNDERSTANDING MATHEMATICS THROUGH RESOLUTION OF PARADOXES

Margo Kondratieva
Memorial University of Newfoundland, Canada

ABSTRACT

Brain-challenging puzzles have attracted people for a very long time. Paradoxes constitute a special type of puzzle aimed to reveal and emphasize an inconsistency or contradiction resulting from some mental experiments in mathematics. Their resolution teaches us to stay alert and be aware of possible flaws of various kinds. Many paradoxes, such as those of Zeno and Russell, greatly influenced the shape of mathematics as we know it today. That suggests a possibility to incorporate the study of paradoxes in standard mathematical courses. But how productive may it be? At which stage of their study will students benefit from being exposed to paradoxes? How one can practically do it in the classroom? This paper is an attempt to address some aspects of these important questions. We discuss the nature and role of paradoxes in the process of understanding, along with potential problems and advantages of their use in study. We give several examples of mathematical paradoxes in both the historical and the classroom context. A short survey results outline an idea of the audience reaction and suggests further directions for research.

Interdisciplinarity, Creativity, and Learning, pages 61–74

61

We conclude that the pedagogical payoff of the use of paradoxes in the class-room is currently underestimated and a comprehensive and consistent study of the impact of paradoxes on learners will allow us to develop a teaching strategy which takes an advantage of the natural curiosity of the mind towards puzzles.

THE ROLE OF PARADOXES IN THE PROCESS OF UNDERSTANDING AND LEARNING

A paradox in the broad sense is a suddenly emerged unexpectedness, a statement or situation that looks unbelievable, wrong, or contradictory. Its presence facilitates the process of understanding things in an attempt to fix an error and to make sense. Somewhat similar we find in the following description of a metaphorical language perception (Carson, 2000).

> In his discussion of metaphor in the *Rhetoric*/Aristotle says there are 3 kinds of words./ "Strange, ordinary and metaphorical./ Strange words simply puzzle us;/ ordinary words convey what we know already; it is from metaphor that we get hold of something new & fresh"/ (*Rhetoric*, 1410b10–13)./ In what does the freshness of metaphor consist?/ Aristotle says that metaphor causes the mind to experience itself/ in the act of making a mistake./ He pictures the mind moving along a plane surface/ of ordinary language/when suddenly/ that surface breaks or complicates./ Unexpectedness emerges./ At first it looks odd, contradictory or wrong./ Then it make sense./ And at this moment, according to Aristotle,/ the mind turns to itself and says:/ "How true, and yet I mistook it!"/ From the true mistake of metaphor a lesson can be learned./ Not only that things are other than they seem,/ and so we mistake them,/ but that such mistakenness is valuable./ Hold into it, Aristotle says,/ there is much to be seen and felt here./ Metaphors teach the mind/ to enjoy error/ and to learn/ from the juxtaposition of *what is* and *what is not* the case.

The metaphor, as described by Aristotle, presents a paradox at the initial moment of establishing non-obvious relations. Later, the paradox becomes resolved by confirming or accepting the identification. In a multi-layer metaphor, another kind of paradox may occur, when an unacceptable conclusion emerges from parts that make sense as they are, but the integrated whole does not. That forces one to review the entire way of reasoning and give up one of the parts.

If we accept the view of Lakoff and Nunez (2000) that mathematics has a metaphorical nature, we have to admit, following Aristotle, that in the whole history of mathematics, in the course of its development, the mind has been permanently and constantly exercised in the resolution of para-doxes of various kinds and levels. Over the centuries, this resulted in the mathematical theory with its definitions, rules and statements, the justification and meaning of which can be fully understood only via personal experience of the resolution of the paradoxes for yourself. In a way, the event of each discovery repeats at the level of an individual learner who is trying to understand the essence of a statement. Having said that, I do not suggest a rediscovery by a learner of the entire theory from scratch, but rather a rediscovery of the meaning of the existing theory. As an example of learning from a paradox, let us consider the following algebraic derivation.

P1: *Assume that A = B. Multiply both sides by A to get A * A = B * A. Subtract B * B from both sides: A * A − B * B = B * A − B * B. Factor both sides (A + B)(A − B) = B(A − B). Divide both sides by (A − B): A + B = B. Recall that A = B, so B + B = B. Therefore, 2 = 1.*

The arrival at the contradictory statement makes one analyze the derivation line by line and eventually realize why *division by zero* is prohibited in algebraic manipulations, as opposed to just blindly accepting the rule from a book or teacher.

Knowing rules is very important for doing mathematics. But mathematical rules cannot substitute for understanding. First of all, according to Godel's discovery made less than a century ago, a complete system of formal rules is simply impossible. This discovery ruined Hilbert's agenda for *complete axiomatization* of the theory with further deductions using a perfect artificial (not appealing to intuition) language of reasoning as it was seemingly suggested by the ancient Greeks' geometry and philosophy of Euclid and Plato. On a positive note, the discovery by Godel tells us that the door is open for insight, creativity, informal approaches and imagination in mathematical thinking (see Penrose,1994, p. 72). To survive and be transmitted from one generation to another, mathematics needs an interpretive, active mind, not just a structured storage of its facts. The understanding of mathematics seems to appear through this interpretation of the real imperfect world in pure abstract terms, through a desire to capture in symbols the things we apparently know from our living experience (like time, space etc.)

The role and place of paradoxes in the process of cognitive development can be identified with the help of Piaget's concepts of *equilibration* and *reflective abstraction*, which rest on the Kantian epistemological proposition that *a knower constructs their knowledge of the world.*

Implicit in Piaget's analysis is the idea that knowledge and understanding are acquired only as the epistemic subject applies its existing cognitive structures to that of which it becomes aware (cognitive aliment) . . . Equilibration refers to a series of cognitive actions performed by a knower seeking to understand cognitive aliment, which is experienced as novel, resistant, perturbing, dis-equilibrating. This experience of dis-equilibration motivates the knower to attempt to re-equilibrate . . . The most interesting form of equilibration is that in which particular cognitive structures re-equilibrate to a disturbance by undergoing a greater or lesser degree of re-construction, a process known as reflective abstraction. We would agree that in those cases in which successful learning occurs, reflective abstraction has taken place. (Dubinsky & Lewin, 1986)

Since a paradox (meaning in Greek "beyond belief") provides a disequilibration, it makes the subject realize the need to re-equilibrate. In this quality, paradoxes are valuable components which stimulate learning and discovery processes by means of restructuring existing schemata of the learner (Rumelhart and Norman, 1978).

On the historical scale one can, for example, argue that paradoxes on infinity, attributed to Zeno (5th century B.C.), eventually led to the discovery of calculus by Newton and Leibnitz in the 17th century, contributing to the notion of the sum of infinite series and the notion of limit. For instance, the fact that the sum of the geometric series $1/2 + 1/4 + 1/8 + \ldots$ is equal to 1 can be traced back to one of the most famous Zeno paradoxes on the subject of motion.

P2: *There is no motion, because that which is moved must arrive at the middle before it arrives at the end, and so on ad infinitum.*

Zeno concludes that the runner can never reach his destination because he must first run one-half of the distance, then an additional one-fourth, then an additional one-eighth, etc, always remaining short of his goal. And here is Aristotle's critique: "Zeno's argument makes a false assumption when it asserts that it is impossible to traverse an infinite number of positions in a finite time." The modern mathematical idea of the limit, and thus, whole mathematical analysis rests on such considerations. Note however that the idea of the limit was successfully used already by Archimedes who calculated the area of a circle by means of a sequence of inscribed regular polygons.

In a classroom perspective, a student, puzzled by a paradox, welcomes help and attends to the hints and related explanations to a greater degree, so the lesson becomes more productive and the theory more meaningful for such a student. The amount of hints and help critically depends on the student's background and abilities. "Students with talent for mathematics, once dis-equilibrated, do perform reflective abstraction on their own, spontaneously. Most students need help with this step" (Dubinsky and Lewin, 1986, p. 90; Sriraman, 2004).

On the other extreme, "A student may persist in applying an incorrect understanding of a concept, even after a teacher pointed out the error."(ibit, p. 88). If a teacher can come up with an example of how the error leads to a paradox (e.g., **P1** above), that can help to correct the student's behaviour.

The case of belief persistence is known to be difficult for pedagogical treatments due to the psychological nature of it: people continue to believe in a claim even after the basis for the claim has been discredited. Nevertheless, a paradoxical consequence may initiate some change in the learner's conceptual state, because she now has to live with both—the persistent concept and its nonsense conclusion.

Resolution of paradoxes is always an effort to overcome one's *mental set* or *functional fixedness* the state when problem solvers rely too heavily on their previous concepts, expectations, strategies and approaches. One needs a fresh look, a new concept, a different method to show that either the contradiction is only apparent, or the paradox rests on invalid or unreasonable grounds. It is crucial whether one believes that the nature is paradoxical[1] or that any paradox is such only due to restrictive frameworks in which one is trying to make sense of it. The later point assumes that the paradox can and should be resolved by embedding it in a wider context and possibly rethinking the meaning of the situation.

There is one more thing we wish to mention following Danesi (2002, p. 227).

> The suspense that accompanies an attempt to find a solution to a challenging puzzle, or anxiety that develops from not finding one right away, is a significant part of what makes the puzzle so fascinating and engaging...The peculiar kind of pleasure puzzles produce can be called an aesthetics of mind...They can never be characterized as sad or happy; they can only be called as ingenious or clever...which produce a form of pleasure nonetheless...Puzzles with simple yet elegant solution, or puzzles that hide a non-obvious principle, have a higher aesthetic index. The aesthetic index is also high when demonstration produces a paradoxical result.

DIFFERENT TYPES OF EXAMPLES AND PARADOXES[2]

One finds puzzles already in the Egyptian Rhind Papyrus written around 1650 B.C. The earliest known mathematical paradox is the *Cretan Paradox*, dating back to the sixth century B.C. attributed to the semi-mythical Cretan poet Epimenides. The poet said: "All Cretans are liars. "

Assuming that this statement is true one immediately concludes that it is false because it was proclaimed by a Cretan. In a certain interpretation of its translation from Greek the assumption that the statement is false also becomes problematic. A refined version of this paradox is attributed to Eubulides, who lived in the fourth century B.C.

P3: *A man says that he is lying. Is what he says true or false?*

The 20th century restatement of the question, the *Barber Paradox*, was given by the English philosopher and logician Bertrand Russell in 1903.

P4: *In a town, the barber shaves all and only those who do not shave them-selves. Does the barber shave himself?*

Such sentences can be classified as *Semantic paradoxes*. A statement and its negation are asserted simultaneously: an assignment of a truth value to the statement ultimately implies the assignment of a different value. Other examples of semantic paradoxes are:

P5: *This statement is false.*

P6: *Every rule has an exception.*

Main feature of these logical paradoxes is that they are a self-referential. "Drawing Hands," composed by M. C. Escher in 1948, provides a visual anal-ogy for this type of paradox.

Paradoxes involving limiting processes and infinite sets constitute another large category, which includes the Zeno paradox (P2) mentioned above. Mental games with *infinity* are proven to be dangerous and often lead to surprises.

P7: *Galileo's paradox: there are as many squares, $n*n$, as natural numbers, n, but there are also natural numbers which are not squares: 2, 3, 5, . . .*

P8: *An infinitely long Gabriel's horn resulting from the revolution of the hy-perbola ($y = 1/x$, $x > 1$) about the x-axis, has finite volume and yet an infinite surface area.*

Fractals are objects resulting from an infinite self-similar reproduction; they have some *strange* properties.

P9: *The fractal called Koch snowflake has finite area and infinite perimeter at the same time. Another fractal called Sierpinsky carpet has zero area and yet has a non-empty interior.*

There are many *Paradoxes resulting from flawed reasoning, arithmetic error or faulty logic*. One of them, a proof that "All horses are the same colour," is analyzed in Section 4. The following *Missing Dollar Paradox* may even be considered as an arithmetic joke, because it has so simple solution.

P10: *Three travelling salesmen break down and are forced to spend the night at a small town inn. They go in and the innkeeper tells them, "The rooms are $30." Each man pays $10 and they go up to the room. The husband of the innkeeper says to her, "Did you charge them the full amount? Why not give* *them five bucks back since their car is broken and they hadn't planned to stay here."* She then brings the men five $1 bills and each man takes one while the other $2 rest on the table. Originally each man paid ten dollars (10 * 3 = 30); now each man has paid nine dollars (9 * 3 = 27) and there are $2 sitting on the counter (27 + 2 = 29). The last dollar had disappeared.

*A simple solution can be offered for this paradox: the amount that the salesmen spent is (10 – 1) * 3 = 27*; amount that the innkeeper received is (30 – 5) + 2 = 27; there is no missing dollar.

However, the next puzzle, known as an Unexpected Exam Paradox is more elegant.

P11: *A professor announces that he is giving an examination some day next week, and the exact day of the examination will surprise his students. One student reasons about the professor's statement as follows. Suppose the exam is on Friday (the last day of the week). Then on Thursday night, the students will know that the exam will be on the next day, so there is no surprise. Hence the exam will NOT be on Friday. Now suppose the exam is on Thursday. Then on Wednesday night, they will all know that the exam is on the next day: again there is no surprise and therefore the exam will not be on Thursday. The same argument applies to Wednesday, Tuesday and finally Monday, leaving the conclusion that there will not be any exam in the following week. The sequel to this story is that when this student receives the exam paper, he is very surprised.*

True statements which contradict common intuition are often regarded as paradoxes. One needs to overcome his/her *epistemological obstacles* (Sierpinska,1994) in order to fully resolve such paradoxes. For instance, one can never *come to believe* that $1 = 0.999999\ldots$, even after learning the way the statement is understood. It is the same as Zeno thought about his runner. Many probabilistic statements are found to be counter-intuitive.

P12: *Birthday paradox: if there are 23 or more students in a class then there is a chance of more than 50% that at least two of them will have the same birthday.*

P13: *Monty Hall Paradox: Suppose you're on a game show, and you are given the choice of three doors: behind one door is a car; behind the others, goats. You pick a door, say No. 1, and the host, who knows what is behind the doors, opens another door, say No. 3, which has a goat. He then says to you, "Do you want to pick door No. 2?" Is it to your advantage to switch your choice? Answer: Yes, a player who has a policy of always switching will win the car on average two times out of the three.*

Paradoxes often result from an intentional error (algebraic, geometrical, logical) in a *model problem.* They teach us how to avoid similar errors in *real problem* solving. They warn us, for instance, that a proof by picture may be not reliable, that a word description may describe nothing, that formal manipulations may lead to an absurdness, and that something counter-intuitive may still be true. Paradoxes teach us to be sensitive and alert at every step we make in our derivations.

TEACHING WITH PARADOXES: THEORY AND PRACTICE

There are two stages in the process of paradox understanding. First is the recognition of its ingredients—words, terms, graphs etc., as well as the underlying assumptions. Second is the realization that even if each part does make sense, the statement as a whole does not work. In order that the second stage produce the effect of confrontation (dis-equilibration), the first stage should be a simple cognitive act, presenting no difficulties, troubles or doubts. In other words, the learner should operate in a familiar clear environment to be able to analyse the situation and trace the fallacy.

Following Dubinsky and Lewin (1986) we will distinguish alpha-, beta- and gamma- types of behaviour. Alpha-behaviour is the integration of a novelty in an unstable or inadequate way: "students thought that they understood and mastered a particular concept that they, in fact, had not." In contrast, beta-behaviour signifies the re-organization of the cognitive system of the learner and the accommodating of the novel aliment within the reconstructed domain. A gamma-behaviour consists in an accommodation of a novelty without a preceding reconstruction of the cognitive system.

When Does a Student Not Learn from Paradoxes?

Many students have developed (over years in high school) a habit of pseudo-learning, mostly in a declarative mode, using explicit rules, given formulas, memorized facts. This often leads to the alpha-behaviour, which the learner has a tendency to substitute for an understanding. Such behaviour often results from a state of learners' minds in which a formal definition of a concept differs from its internal cognitive representations (Tall and Vinner, 1981). Consequently, a learner experiences confusion and paradoxes resulting from her own misconceptions and psychological obstacles to accept certain mathematical statements. At this point there is no room for critical learning of new material. Therefore such a learner is reluctant to be exposed to further contradictions. She does not have a way or habit to resolve them. For such a student, often individual assistance is required before she starts to accept the very idea of creative thinking.

When Does a Student Successfully Learn from Paradoxes?

The above classification suggests that when a paradox is introduced, a learner is expected to exhibit a gamma-behaviour at the first stage, in other words, her cognitive system should be sufficiently rich to accommodate the parts by themselves. The second stage, the realization of the contradiction, then will initiate the beta-behaviour targeting for the resolution. Such an interpretation is possible in a dialectic approach in which the learner's behaviour type is viewed relative to the task, experiencing changes due to the task performed. The beta-behaviour, being "the paradigm case of successful learning," is identified not only as an ability to produce a correct answer by reasoning, but as a readiness to review, analyse, and re-construct the existing cognitive system, if required. The ability to work with both concept image and concept definition, to create multi-linked representations, structure, and to fit ideas into a big picture are characteristics of advanced mathematical thinking. This type of thinking may be present at all grade levels and my point is that it is a prerequisite for beneficial learning through resolution of paradoxes.

During resolution of a paradox, a learner makes a revision of her understanding of the ingredients: terms, graphs, logical implications, and hidden assumption. The contradiction she faces is her motivation to localize the source and reason for it. Everyone who manages to understand the essence of a paradox wants to know a resolution. This is a natural curiosity of mind. That curiosity forces the mind to make guesses. But if all attempts are unsuccessful and the ideas seems to be exhausted, a certain reluctance to proceed may be observed. (The reluctance is a sort of defensive reaction of the learner, unable to resolve the situation immediately. She may even convince herself that it is not important to complete the task.) Nevertheless, the process of the contradiction resolution still goes on internally, subconsciously, and the mind stays open for a possible hint from the environment. An it *aha-moment* can occur any time.

This state can be effectively used for teaching mathematical concepts related to the essence of the paradox. A desirable outcome is to make the learner to apply the concepts to resolve the paradox. A whole lesson can be designed to lead the discoverer to her goal. An example of such practice is given by Lahme and McDonald (2006), when the entire course was built around discussions of paradoxes.

My experience of talking about paradoxes with university students has significant variations. There are students who definitely and undoubtedly learn a great deal from thinking about contradictions, students who this way invent and rediscover for themselves important mathematical constructions and notions. There are students who understand the essence of a puzzling statement, who become curious about the true reasons for it and appreciative of its various interpretations and explorations. Not being exposed to paradoxes would be to disadvantage of such students.

There is another group of students, for whom the craft of learning from paradoxes is not fully accessible. They still can be skillful students, hard working, good solvers of standard tests, interested in mathematics, enjoying simple puzzles, but surprisingly helpless even with problems like the Missing Dollar Paradox (P10). Is it a lack of critical thinking? Not understanding when formal proof becomes a fallacy? Inability to decompose a long argument into more elementary statements? Anyhow, such students need special assistance to be able to benefit from paradoxes.

There is a third group of students who do not benefit from being directly exposed to paradoxes, they are getting misled, confused, frustrated. Their presence almost make a case against using paradoxes in teaching.

In order to estimate the relative sizes of those groups and the effect of working with paradoxes a survey study has been conducted among second and third year undergraduate students. The students were asked to resolve few paradoxes, including (P1), and then answer questions about their experience with paradoxes. It would be worth mentioning that among the group of students I interviewed, 35% take math courses because it is required in their program, 32% study mathematics because they find it to be interesting, 38%—because they believe that math is useful, and 15% admitted that they always liked mathematics. In this and following responses some students gave more then one answer, thus the total percentage may exceed 100.

It was observed that 62% of the students were able to resolve and explain paradox P1, identifying the wrong step and the forbidden operation; 15% of the students were able to come up with some explanations, but did not give a clear resolution. For example, "Once you eliminate $(A - B)$ on both sides you are left with $A + B = B$, so something wrong is here." Another student wrote: "The only possible solution is $A = B = 0$. Then $B + B = B$ makes sense." The remaining 23% of the students could not say anything relevant, except "2 is not equal 1," or "This can't be right," or "Magic!"

The students were also asked whether or not there is a notion or fact in mathematics which they regard as counter-intuitive. Most of the students did not find anything of that nature, although a few examples were listed: "any number to the zero power is one," "zero factorial is one," "the notion of infinity," "harmonic series diverges," "the sum of the inverse squares of all natural numbers is related to the number pi." One student gave a long list of paradoxes and concluded that he would be willing to take a course about paradoxes.

Finally, the questions of the survey were answered in the following way:

1. Resolution of paradoxes is challenging 30%, interesting 65%, emotional 30%. No one said that it is boring. One student said that it can be annoying.
2. Sixty six percent of the students were in favor of discussion of paradoxes in class, and 34% did not welcome this opportunity. Forty four percent of the students thought that it would improve their understanding, 26% felt that it would be fun, 12% wished to learn about historical development of mathematics. On the other hand, 12% admitted that they have enough confusion about math already, and 32% felt that they would rather spend time on regular problems. Nevertheless, some would like to discuss paradoxes if they are relevant to the area of their study.

These observations suggest that an instructor must be careful and sensitive while implementing paradoxes into a teaching process. It is a challenge and a risk, but it can be a valuable exercise in the creative thinking and problem solving orientations, if used deliberately and cleverly. I summarize some recommendations suggested by this study.

- Build a collection of paradoxes and include them in the teaching portfolio as a specific challenging activity in the classroom (new paradoxes are supplied by periodic editions e.g., *College Mathematics Journal*).
- Teach paradoxes to illustrate a specific point. State the point clearly. Make sure that the essence of paradox is understood in the context of the material being studied.
- Allow the students some time to think on their own.
- Engage the students into a discussion. Make them develop their conclusions, leading to a resolution.
- State a resolution clearly.
- Remember that the goal is to dis-equilibrate the learner in an aesthetically valuable manner, and then make use of her natural curiosity.

Will paradoxes serve that purpose better then other teaching approaches? It may well depend on the instructor's attitude towards paradoxes. But the same is true about any other tools and techniques, and there is no unique best one. Successful teaching is a mixture of many approaches, and we hope to bring paradoxes to the teachers' attention as one of them.

TEACHING WITH PARADOXES: AN OBSERVATION FROM A CLASSROOM

After teaching the method of mathematical induction and giving several examples of its use, the teacher announces that she is going to prove that *All horses are the same colour.* The goal is to make the students discover the flaw in the proof.

Teacher: We use the principle of mathematical induction. As the basis case, we note that in a set containing a single horse, all horses are clearly the same colour. Now assume the validity of the statement for all sets of at most N horses. Consider a set of N + 1 horses. Take two of its subsets: all horses from the first to the Nth, and all the horses from the second to the (N + 1)th. Both subsets contain N horses, and thus all horses in each of them are the same colour by the induction assumption. Because the two subsets have a non-empty intersection the horses in the first subset are same colour as those in the second. Thus the union contains N + 1 horses of the same colour. By the principle of induction, we have established that all horses are the same colour.

Student A: This is clearly impossible.

Student B: But she proved it!

Student A: *(repeats the argument of the implication from N to N + 1 set of horses, confirms it, gets puzzled)* I do not understand it anymore...

Teacher: Let's try some values, say N = 10. (*Intentionally takes bigger number.*)

Student A: Any 10 horses are of the same colour by the assumption, and any 11 horses can be viewed as a union of two subsets: from the first to the tenth and from the second to the eleventh. The subsets intersect, thus any eleven horses are the same colour. There is no way to be mistaken about this.

Teacher: And it works for N greater that 10.

Student B: Yah, I never fully trusted this method!

Student A: Hold on, let me check smaller N. Let me check all over again. One horse is of its colour. True. But two are not. They may be not the same colour. Aha, the induction is gonna get stuck here. It does not work for N = 1! Here is the hole in the proof!

Teacher: Please elaborate.

Student A: You made the implicit assumption about a non-empty intersection of the two subsets of horses to which you apply the induction step, but this is not so if each subset contains one horse and the union two. So the induction step does not work for N = 1, and thus the conclusion is wrong!

Student B: So the falling chain of dominos has a gap at the very beginning! *(Referring to the visual analogy the teacher used to describe the induction as the chain of dominos each forcing to fall the next one.)*

Teacher: Exactly! Thus this is not a paradox, but merely the result of flawed reasoning; it exposes the pitfalls arising from failure to consider special cases for which a general statement may be false.

A few remarks are in order. Both students are puzzled by the proof, but Student A is clearly the one who found the flaw, whereas Student B is more like not analyzing the problem, but making supportive claims from what he remembers. At some point he even doubts the method's validity, because he did not develop trust in it yet. It is likely that student A would find the resolution on his own, but it is student B who needs the discussion and help.

CONCLUSION

In this article we attempted to show that mathematical paradoxes carry a special challenge, a hidden message, thinking of which provides the learner with an opportunity to refine their incomplete understanding. The degree of incompleteness of knowledge defines how productive such an experience can be and how much assistance is required. At the present time, the practice of using paradoxes in mathematical courses is not common, despite the fact that the subject intrigues many learners. Therefore, of great interest to cognitive sciences would be a more detailed investigation of the influence of instructor's extensive and creative use of mathematical paradoxes on students' attitudes and performance. Such research is not only of academic interest but is called for by pedagogical practices. It would allow building a portfolio of challenging activities incorporating paradoxes in a way the learners would fully benefit from them.

REFERENCES

Carson, A.(2000). Essay on what I think about most, in *Men in Off Hours*, New York: Alfred A. Knopf.

Danesi, M. (2002). *The Puzzle Instinct*, Indiana University Press.

Dubinsky, E., & Lewin, P. (1986) Reflective abstraction and mathematics education: the genetic decomposition of induction and compactness. *The Journal of Mathematical Behaviour*, 5, 55–92

Lahme, B, McDonald, E. (2006). Infinity and Beyond -A Mathematics Class for Life Long Learning. *Focus*, 26(1), 16–18.

Lakoff, G., Nuvez, R.E. (2000). *Where mathematics comes from*. New York: Basic Books.

Olin, D. (2003). *Paradox*. McGill-Queen's University Press.

Penrose, R. (1994). *Shadows of the mind*. Oxford University Press.

Rumelhart, D.E., Norman, D.A. (1978). Accretion, tuning and restructuring: three modes of learning. In *Semantic factors of cognition* (pp. 37–52). Hillsdale, NJ: Lawrence Erlbaum Associates.

Sierpinska, A.(1994). *Understanding in mathematics*. London: Falmer.

Sriraman,B. (2004). Reflective abstraction, Uniframes and the formulation of generalizations. *The Journal of Mathematical Behavior*, 23(2), 205–222.

Tall, D., & Vinner, S. (1981). Concept image and concept definition in mathematics with particular reference to limit and continuity. *Educational Studies in Mathematics*, 12,151–169.

NOTES

1. See e.g., works by Graham Prist on paraconsistency and dialetheic logic, and more discussion in (Olin , 2003).
2. Most of the paradoxes given in this paper are well known. Their formulations and further discussion can be found at Wikipedia (http://en.wikipedia.org)

ACKNOWLEDGEMENT

Reprint of Kondratieva, M. (2007). Understanding Mathematics through Resolution of Paradoxes. In B. Sriraman (Ed.). Perspectives on Talent Development in Mathematics. *Mediterranean Journal for Research in Mathematics Education*. 6(1 & 2), 127–138. ©Margo Kondratieva & Bharath Sriraman.

CHAPTER 7

MATHEMATICAL PARADOXES AS PATHWAYS INTO BELIEFS AND POLYMATHY

Bharath Sriraman
The University of Montana

ABSTRACT

This paper addresses the role of mathematical paradoxes in fostering poly-mathy among pre-service elementary teachers. The results of a 3-year study with 120 students are reported with implications for mathematics pre-service education as well as interdisciplinary education. A hermeneutic-phenomeno-logical approach is used to recreate the emotions, voices and struggles of students as they tried to unravel Russell's paradox presented in its linguistic form. Based on the gathered evidence some arguments are made for the benefits and dangers in the use of paradoxes in mathematics pre-service education to foster polymathy, change beliefs, discover structures and open new avenues for interdisciplinary pedagogy.

Interdisciplinarity, Creativity, and Learning, pages 75–93

INTRODUCTION

Elementary set theory serves as the backbone of mathematics content required by prospective elementary school teachers around the world. This is evident in the content standards of numerous curricular documents (e.g., Australian Research Council, 1990; National Council of Teachers of Mathematics, 2000) which call for both a foundational and contextual understanding of the models for the four arithmetic operations $(+, -, \times, \div)$ developed for the natural, whole, rational and real numbers. In addition the use of manipulatives such as Dienes base–10 blocks, Cuisenaire rods etc, greatly facilitate the enact-ion of the elementary arithmetic operations for the particular set under consideration. However the set theoretic and philosophical foundations of these operations are typically thought to be beyond the scope of pre-service education. The two fundamental questions explored in this paper are: (1) How can we facilitate the discovery of the mathematical foundations, paradoxes and structures? and (2) How can deeply rooted beliefs about the nature of mathematics be impacted?

MOTIVATION AND CONCEPTUAL FRAMEWORK

In the United States, teacher professional development programs typically target in-service teachers and use an interventionist attempt to shift their beliefs and practices about the nature of mathematics. A large body of extant research addresses pre-service and practicing school teachers' beliefs and attitudes (e.g., Thompson, 1992) towards mathematics and describes the affective factors (Leder, Pehkonen & Törner, 2002) which influence mathematical understanding (Ball, 1990)and problem solving (Goldin, 2000, 2002). There is also research which addresses limitations of current research approaches to studying teacher beliefs (Leatham, 2006; Wedege & Skott, 2006). Wedege & Skott (2006) argue that the main-stream trend of

> research on belief-practice relationships runs the risk of becoming a self-fulfilling prophecy. It often contains a circular argument of claiming that certain observed mathematical practices are due to beliefs, while at the same time inferring mathematical beliefs from the very same practices. (p. 34)

Similarly Leatham (2006) critiques research on teacher beliefs as assuming that

> teachers can easily articulate their beliefs and that there is a one-to-one correspondence between what teachers state and what researchers think those statements mean. Research conducted under this paradigm often reports inconsistencies between teachers' beliefs and their actions. (p. 91)

Ernest (1989) categorized three philosophies of mathematics, namely the instrumentalist view, the Platonist view, and the problem solving view. The instrumentalist sees mathematics as a collection of facts and procedures which have utility. The Platonist sees mathematics as a static but unified body of knowledge. Mathematics is discovered, not created. The problem solving view looks on mathematics as continually expanding and yet lacking ontological certainty. The problem solving view sees mathematics as a cultural artifact. This implies that what is thought as true today, may not be seen as true tomorrow. (pp. 99–199). Ernest also describes absolutist and fallibilist views of mathematical certainty. The absolutist sees mathematics as completely certain and the fallibilist recognizes that mathematical truth may be challenged and revised (Ernest, 1991, p. 3).

Lerman (1990) recognizes that ones philosophy is related to ones preferred teaching style. The absolutist teacher will prefer a direct teaching style whereas a fallibilist is much more likely to engage in exploratory activities and open-ended problems. In what is now one of the seminal studies in the domain of teacher beliefs, Thompson (1984) studied three junior high teachers, all of whom had different beliefs about the nature of mathematics. The first teacher viewed mathematics as a coherent collection of interrelated concepts and procedures. She regarded mathematics as a subject free of ambiguity and emphasized conceptual development in the students. She would fit Ernest's model as a Platonist. The second teacher had a very different perspective of mathematics. Thompson says her teaching reflected more of a process-oriented approach than a content oriented approach. A view of mathematics as a subject that allows for the discovery of properties and relationships through personal inquiry seemed to underlie her instructional approach. This teacher would fit Ernest's model as person with the problem solving view. The third teacher in Thompson's study saw mathematics as a collection of facts and procedures which help students find the answer. She saw no ambiguity in mathematics. She would fit Ernest's model as a person with an instrumentalist view. Thompson sees at least three distinct ways of viewing mathematics, all of which greatly influence the choice of curriculum and its delivery. Thompson (1992) says that research on teachers cognitions and studies of teachers' conceptions have contributed to a conceptual shift in the field of research on teaching, moving away from a behavioral conception of teaching towards "a conception that takes account of teachers as rational beings" (p. 142). Our understanding of teaching from teachers' perspectives complements our growth of understanding of learning from learners' perspectives, which in turn, enriches the idea of schooling as the negotiation of norms, practices and meanings (Cobb,1988). Much earlier, Fenstermacher (1978) predicted that the single most important construct in educational research are beliefs (Pajares 1992). Törner & Sriraman (2007) argue that Thompson's (1982,

1992) theory to explain teacher's actions in a mathematics classroom based on their beliefs about mathematics is one instance of the development of a local philosophy based on problematizing research in the domain of beliefs. Thompson (1992) wrote:

> I think we will get further evidence on the role of teachers' views of mathematics when we go into more detail and investigate their understanding of different domains of mathematics, of specific components such as the meaning of mathematical concepts, proof, definition, theorem, conjecture, variable, symbols, rule, formula, axiom, problem, problem solving, application, model, computation, graphical representation, visualization, metaphor,etc., both with respect to the various sub-domains of mathematics as well as in a more general sense. (p. 142)

Today we usually speak of teachers' beliefs, which are generally formulated as "views about mathematics" (e.g., Grigutsch 1996; Pajares 1992). It is assumed that different beliefs about mathematics have different associated philosophies and/or epistemologies (Törner, 2002). Amidst all this important research which increases our understanding of teacher beliefs, in lieu of Thompson's call, one is left wondering about the dearth of studies (recent or otherwise) in describing or analyzing prospective elementary school teachers understanding of the foundational (set-theoretic) concepts of the mathematics they are exposed to. Could it be that the aftermath of New Math has had an "affective" impact (pun-intended) on the focus of mathematics education researchers engaged in teacher education and led to less emphasis of this particular mathematics content. For the author it is important that we attend to the significant role that tasks may have in teacher education, particularly tasks that foster interdisciplinary thinking as well as tasks that shake the dominant views of prospective teachers on the nature of mathematics. In the remainder of this paper the author will report on the use of a set-theoretic task to help prospective elementary school teachers understand and discover paradoxes and structures and foster polymathy. Based on the findings, limitations and dangers in such tasks are also outlined.

POLYMATHY, PARADOXES, AND PHILOSOPHY

The term polymath is in fact quite old and synonymous with the German term "Renaissance-mensch." Although this term occurs abundantly in the literature in the humanities, relatively few (if any) attempts have been made to isolate the qualitative aspects of thinking that adequately describe this term. Most cognitive theorists believe that skills are domain specific and typically non-transferable across domains. This implicitly assumes that "skills" are that which one learns as a student within a particular discipline. However such an assumption begs the question as to why polymathy occurs in the first place? (Sriraman & Dahl, 2007).

Root-Bernstein (1989,1996, 2000, 2001, 2003) has been instrumental in rekindling an interest in mainstream psychology in a systematic investigation of polymathy. That is the study of individuals, both historical and contemporary, and their interdisciplinary thinking traits which enabled them to contribute to a variety of disciplines. Common thinking traits of the thousands of polymaths (historical and contemporary) as analyzed by Root-Bernstein, Robert Sternberg , Dean Simonton and many others are: (1) Visual geometric thinking and/or thinking in terms of geometric principles, (2) Frequent shifts in perspective, (3) thinking in analogies, (4) Nepistemological awareness (that is, an awareness of domain limitations), (5) Interest in investigating paradoxes (which often reveal interplay between language, mathematics and philosophy), (6) Belief in Occam's Razor [Simple ideas are preferable to complicated ones], (7) acknowledgment of Serendipity and the role of chance, and (8) the drive to influence the Agenda of the times. Although the prodigious writings of these researchers (Root-Bernstein, 1989, 1996, 2000,2001, 2003; Sternberg et al., 2004) typically involve eminent individuals, it has been found that polymathy as a thinking trait occurs frequently in non-eminent samples (such as high school students) when presented with the opportunities to engage in transdisciplinary behavior. In particular the use of unsolved classical problems, paradoxes and mathematics literature has been found to be particularly effective in fostering inter-disciplinary thinking (Sriraman, 2003a, 2003b, 2004, 2005).

METHODOLOGY

In order to investigate whether prospective elementary school mathematics teachers display some of the thinking traits of polymaths the author conducted a 3 year study with approximately 120 prospective elementary school mathematics teachers in the 2002–2005 time period. These pre-service teachers were enrolled in the mathematics content sequence for elementary teachers at a large university in the western United States. This two semester content sequence is the only required mathematics content for these prospective teachers due to the particular state legislations in this region. The author was also the instructor of these students. Journal reflective writing was an integral part of this course.

During the course of the semester, among the various tasks assigned to the student was the following.

The Task

The town barber shaves all those males, and only those males, who do not shave themselves. Assuming the barber is a male who shaves, who shaves the barber? Explain in your own words what this question is asking you? When you construct your response to the question, please justify using clear language why you think your answer is valid? If you are unable to answer the question who shaves the barber, again justify using clear language why you think the question cannot be answered.

This task is the well known linguistic version of Russell's paradox, appropriately called the Barber Paradox. The question as formulated here was read out several times in the class in order to clarify what it was asking. Students were given about 10 days to construct a written response to this task. The purpose of this task was to investigate whether students with no prior exposure to the paradox would be able to decipher the contradictions in the linguistic version of Russell's paradox, and whether they would be able to then construct their own set theoretic (mathematical) version of the paradox. All the students were also asked to complete the following affective tasks parallel to the mathematical task. The students were also requested not to consult the worldwide web in search of a solution.

- Write one paragraph (200–300 words) about your impressions of a given question after you have read it, while tackling it (if possible), and after you've finished it.
- In particular record things such as:
- The immediate feeling/mood about the question (confidence, in confidence, ambivalence, happiness, tenseness etc)
- After you've finished the question record the feeling/mood about the question (if you are confident about your solution; why you are confident? Are you satisfied/unsatisfied? Are you elated/not elated? Are you frustrated? If so why?
- Did you refer to the book, notes? Did you spend a lot of time thinking about what you were doing? Or was the solution procedural (and you simply went through the motions so to speak)
- Was the question difficult, if so why? If not, why not?
- Did you experience any sense of beauty in the question and/or your solution?

Data Collection and Analysis

Out of the 120 students, 52 students were able to unravel the paradox, i.e., understand and explain the contradiction in their writings, 40 students believed there was no contradiction (i.e., they answered that the barber shaved himself), and 28 students gave up on the problem but completed the affective portion. Over the 3-year period, in addition to the written journal responses of the students to the aforementioned tasks, the author interviewed 20 students from the 120 students who were representative of the larger sample.

Of these 10 students had successfully unraveled the paradox (out of the 52), and the remaining 10 were unable to unravel it. 6 of these students came from the subset of 40 who saw no contradiction, and the remaining represented those who gave up on the problem. These students were purposefully selected on the basis of whether they were able to unravel the paradox and formulate its set theoretic or mathematical equivalent and those that were unsuccessful in their attempts to do so. It should be noted that all students were given full credit for the assignment irrespective of whether they were successful or not. The written artifacts (students journal writing/solution and affective responses), and interview data were analyzed using a phenomenological–hermeneutic approach (Merleau-Ponty, 1962; Romme & Escher, 1993) with the purpose of re-creating the voice of the students. Phenomenology has its roots in the philosophical work of Husserl and Heidegger, which was extended into a theory of embodiment by Merleau-Ponty in order to counter reductionism, dualism and to capture the totality of human experience.

During the interview students re-voiced their experience of unraveling the paradox. The author simply sought clarifications on the written solution, their affective responses and asked students for general comments on the nature of the problem and their struggles with it. This led to questions on their beliefs about mathematics. There was no pre-set direction or protocol in which the interview was made to progress. Each interview lasted approximately 60 minutes. The individual transcripts and the author's interpretation of student voices, particularly their self-reported affective and polymathic experiences was discussed with each student to ensure validity and reliability. Student journal writings were coded several times to categorize and determine the affective mood self reported by the students. Similarly the interview transcripts were also coded for affective categories and to determine consistency in the self reported voices. In addition, the constant comparative method from grounded theory was applied for the purposes of triangulating the categories which emerged from the phenomenological approach (e.g., Annells, 2006). Finally, in all classes a de-briefing session occurred in which various students presented their views on the problem and their solutions with a discussion of the contradiction.

RESULTS

Examples of affective responses, student solutions, and interview transcripts are now presented in a phenomenological style, which recreate student voices as they struggled through this paradox. Contrasting voices are presented and are representative of the 20 students who participated in the interviews.

Voices

Note the following abbreviations:

JWAV   Journal writing affective voice (response);
JWSV   Journal writing solution voice (response);
IVS   Interview voice student;
IVA   Interview voice author

JWAV1: Started the question in class (10/19) after you (author) read it. Stopped occasionally for work and classes, finished it 10/25 at 12.01pm. The question is very confusing and I feel very anxious about this question. Why would this question be asked in a math class? Don't get me wrong, I like to think about questions like these, but they are difficult and time consuming. I'm frustrated... [w]hat a strange question.

JWAV2: I painfully came to the conclusion that the question was answerable through reading the question several times and thinking about it for days. It is a beautiful paradox, if one thing were changed in the question I feel there would be a definite answer. The English in it is perfect.

JWSR1: The question is asking who shaves the barber? However the barber is a male and he shaves only those males, who do not shave themselves. Thus he cannot shave himself because he only shaves the males that do not shave themselves... [B]ut the barber only shaves those who do not shave themselves... [T]his is a paradox, he cannot be the barber and shave himself, and he cannot shave himself and be the barber.

JWSR2: The barber shaves himself. I justify this firstly by the opening sentence "who do not shave themselves" which implies there are those who do shave themselves. For example a mechanic will work on all of other peoples cars in town but if his car has a problem, he would work on it!

JWAV3: My immediate opinion was that the answer would be "yes." I didn't really feel anything when doing this problem other than that the answer was obvious.

JWAV4: This question was the death of me ... [I] was more than upset with you. Thinking about this question made me so frustrated that I stopped and decided not to waste my time on it. I found this question to have not any beauty in it, all it caused me was a lot of stress and discomfort in my life. It is impossible to put any mathematical equation into it.

The affective response of frustration and curiosity as well as a sense of beauty in the problem was found predominantly in the group that unraveled the paradox (group 1). Students in this group voiced high levels of frustration, anger, curiosity and beauty as well as reported a sense of accomplishment in unraveling the paradox.

In contrast, in the groups that saw no contraction (group 2) and the group that gave up on the problem (group 3), although the levels of frustration were comparably high, the proportion of students that became curious or saw some beauty in the problem was considerably lower. Approximately half of the students in group1 and group 3 became angry while attempting this problem, however nearly 80% of group 1 students reported curiosity whereas only 10% of group 3 students experienced a sense of curiosity. The proportion of self reported experience of "sadness" was considerably higher in group 3 in comparison to the other groups. Group 1 reported the highest levels of accomplishment. Table 7.1 gives a summary of the coded affective voices/responses extracted from student journal writings and interview transcripts.

TABLE 7.1 Affective Traits in Student Groups

Affective mood / Student groups	Frustration	Anger	Curiosity	Sense of beauty	Sense of accomplishment	Set-theoretic formulation	Sadness	No affect reported
Group 1 (52) (unraveled paradox)	52	26	41	28	35	16	0	0
Interview Subgroup 1 (10)	10	7	10	4	8	6	0	0
Group 2[a] (40) (Saw no contradiction—claimed the barber shaved himself)	32	11	14	0	3	0	1	8
Interview Subgroup 2 (6)	3	1	1	0	2	0	1	1
Group 3 (28) (Gave up on the problem)	27	13	3	0	0	0	14	1
Interview Subgroup 3 (4)	4	2	1	0	0	0	3	0

[a] Four of the students in group 2 tried to construct an explanation to unravel the paradox after stating that the barber shaved himself (see Figure 7.2).

Student Journal Writings

In this section, sample journal writings with student solutions are presented.

Commentary 1

This solution represents the numerous solutions obtained from group 1 students where the set-theoretic contradiction became clear to the students. It should be noted that students who discovered and formulated the set theoretic version of the solution spent significantly longer periods of time with the problem in comparison to the other groups. It should be noted that only 16 of the 52 students in group 1 were successful in doing this. It is well known in the foundational mathematics literature that Betrand Russell discovered this paradox and communicated it to Gottlob Frege in a letter right when Frege was about to complete his treatise on the foundations of Arithmetic (*Grundlagen der Arithmetik*), which dealt a devastating blow to his work. The standard way of stating the paradox is if we let R be the set of all sets that are not members of themselves. Then R is neither a member of itself nor not a member of itself.

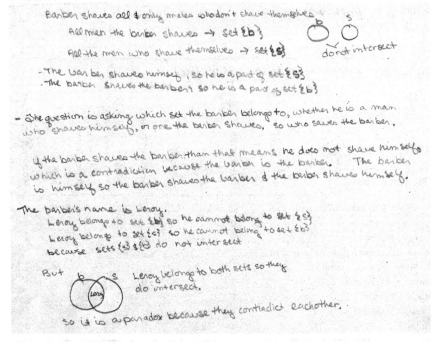

Figure 7.1 Unraveling the set theoretic version of the paradox (Group 1).

> * The question is asking me who shaves the barber if the barber only shaves men who do not shave themselves. This question can not be answered. This is just like the chicken and egg problem of which one came first. People have been trying to answer this question for all of time & never came to an answer. If the barber shaves himself than he does shave people who shave themselves, and if he doesn't shave people who shave themselves than he can't shave himself. Therefore the question can not be answered.

Figure 7.2 Attempting to unravel the paradox linguistically.

Commentary 2

There were 4 students in group 2 who first claimed that the barber shaved himself but then attempted to construct a linguistic explanation like the one shown in Figure 7.2, which were incorrect. Two of these students were interviewed during which they voiced the reasons for doing so, which is reported in a following section of this paper.

Voices 2

In this section, three interview vignettes are presented representing the three groups.

Interview Vignette 1

IVS1: I was tickled by the challenge of this question. I didn't expect to solve a paradox in a math class and had to think a great deal when I was working on other things. I didn't see it as a problem involving sets at least not right away.

IVA1: Did you think it was a mathematical problem?

IVS2: Not really. It made me think of literature or something you would encounter from Zen Buddhism or even Rumi.

a. This question is asking me to explain who shaves the barber (or cuts his hair) if he is the one that is shaving (or cutting) everyone that is male and doesn't shave themselves.

b. If the town barber shaves all those males who do not shave themselves, then the males that shave themselves must shave the barber. If the males that don't shave themselves are Bob, Joe, Bill, Jack, and Frank then the Barber (Tom) shaves them. If the males that shave themselves are Jake, George, Stew and Cody, then they must shave Barber Tom. I think my answer is valid because we know that Joe, Bill, Jack, Bob and Frank don't cut Tom's hair because they don't even cut their own hair. So, the males that do cut their own hair (shave themselves) must cut Tom's hair because there wouldn't be anyone else that could do it. So, Jake, George, Stew or Cody cuts the barbers hair.

Figure 7.3 Unsuccessful solution to the paradox.

Math problems don't make me angry and this one did! I was angry but challenged enough that I didn't let my anger get in the way of finding a solution. I also thought about it in terms of art, like the strange tiling pattern posters we have in class, you know what I mean?

IVA2: Yes, I think you're referring to the Escher posters. How long did you spend on the problem?

IVS3: Yeah… that's the one. Days and days… [b]ut when I saw through it and realized it was a paradox, like you know assuming one thing led to another thing which was contradictory, then I thought it was beautiful. I never imagined saying something like that for a math problem.

IVA4: Why not? So is this a math problem?

IVS4: Umm . . . .in math classes, you like use formulas and equations and stuff and get answers. This one made you think really hard. I wish math were like that. Everything is just black and white in math, atleast this is what we are led to believe.

IVA5: Does this make you change your view of what mathematics is?

IVS5: This problem does, but I don't think we can say that for sure.

IVA6: What if I told you there were many such paradoxes in math? Would that make you see math differently?

IVS6: Sure. Look, we're told all through school that you got to memorize equations and formulas and things like that to do math or like there is a set way to do a problem. But with problems like these you kinda get a little from freedom to think for yourself. So definitely, yeah . . . if we were exposed to more paradoxes like these, we'd change our minds about what math is all about.

Interview Vignette 2

IVA1: You wrote this problem caused the "death" of you? I'm sorry you were so frustrated by the problem

IVS1: Yeah, I was really quite upset with it. I really thought about it for a while and became so confused that I didn't want to deal with it anymore.

IVA2: Would you have approached the problem differently if you were this in a non-math class?

IVS2: I actually did think about my philosophy course when I was reading this problem. But I'm not used to this kind of stuff in a math for elementary teachers class. I did look into the book and it said that there were some paradoxes in math in the first chapter. It was easier to give up than go crazy over the problem. Plus I was really under pressure from work and just didn't have the time to think it through. Sorry.

IVA3: No, no, you don't have to apologize. You simply did what you did. Did you really stop thinking about the problem after you gave up?

IVS3: [silence] Yes and no. I didn't really feel like solving it. Part of the reason might have to do with that I didn't really know what a solution is supposed to look like, you know. I did imagine other situations that could be like the one in the problem but it didn't help any.

IVA4: Do you think paradoxes have a place in mathematics?

IVS4: I don't know. I'm going to do my student teaching soon and I was thinking this might be a fun thing for kids to talk about. I don't know how it would go over with parents if the teacher gave problems like these [laughing].

IVA5: You mean they might get upset?

IVS5: Yeah . . . I mean they'd try to help their kid and get a headache and then blame the teacher for giving this problem. Parents would want the answers to be cut and dry.

Interview Vignette 3

IVA1: I noticed you first created an example with ten people and said it was clear that the barber had to shave himself. Then you expressed doubts after that and said it was like the chicken and the egg problem.

IVS1: My first feeling about the question was complete frustration. I didn't know who or what to relate this question to. How was I supposed to answer the question? So, I did this problem solving thing like making an example. But then I wasn't really using the information given in the problem. Just making stuff up . . . . [s]o I wasn't too happy about it and thought maybe I should do something else. That's when I wrote the stuff about the chicken and the egg problem. Now it seems clear to me that there was a contradiction in the way the problem was given. It was a good brainteaser but very frustrating though.

IVA2: Did you get upset?

IVS2: Honestly, yes. I got pretty worked up. I wasn't satisfied with the solution with people in it. It was all made up. The really upsetting thing was not knowing whether you were right or wrong.

IVA3: Is that how mathematics is supposed to be?

IVS3: [silence]

IVA4: By the way, did the problem make you think of other things?

IVS4: I read and re-read the problem many times and tried to think of it in different ways. I mean I thought of making stuff up like, the barber is married, his wife shaves him. It was tough to stay within the boundaries of the question. It made me think of puzzles or visual tricks that you see in paintings sometimes. Like you see it one way and then you blur your eyes or focus on a different point in the picture and you see something else. You know like those pictures that appear like a duck or a rabbit, or dolphins and people in the same picture. I guess the point of the problem was you couldn't have it both ways. I finally figured out that there were like two sets, one which had all men that shaved themselves, and then another set with all men shaved by the barber, and the barber couldn't belong to either of these sets.

IVA5: Are you happy you figured out the paradox later?

IVS5: I don't know about happy. More like relieved!

Author Voice Over

These three interview vignettes revealed that students were beginning to make connections between mathematics and other domains of inquiry such as philosophy, language and art. The hermeneutic analysis of student journal writings, affective responses and interview transcripts indicated that nearly half of the students (predominantly in group 1) displayed one or more *polymathic* traits when engaged with the paradox. In particular students reported (1) Frequent shifts in perspective (2) thinking with analogies, (3) tendency towards nepistemology i.e., questioning the validity of the question and its place in the domain of mathematics as well as the fallibility of mathematics. The pre-service teachers also reported an increased interest in the place of paradoxes in mathematics, which they had always believed as an infallible or absolutist science. Some of the students began to connect mathematics with language and voiced the need to engage in discourse as opposed to engaging in such an activity solitarily. One of the dominant and consistently heard student voices over the course of three years was of the deeply held dominant belief of mathematics as infallible or absolutist. Table 7.2 gives the polymathic traits displayed by the three groups of students.

IMPLICATIONS AND CONCLUDING POINTS

Polymathy and interdisciplinarity are topics on which one finds scant literature in the field of mathematics education, particularly in domain of pre-service elementary teacher education. Although we live in an age where knowledge is increasingly being integrated in emerging domains such as mathematical genetics; bio-informatics; nanotechnology; modeling; ethics in genetics and medicine; ecology and economics in the age of globalization, the curriculum in most parts of the world is typically administered in discrete packages. The analogy of mice in a maze appropriately characterizes a day in the life of students, with mutually exclusive class periods for

TABLE 7.2 Polymathic Themes Emerging from Journal Writing and Interview Transcript Analysis

Student groups	Shifts in perspective (paradoxes in art and language)	Thinking in analogies	Nepistemological awarness [mathematics as fallible]	Interest in investigating paradoxes	Mathematics as language (need for discourse)	Mathematics as philosophy
Group 1 (52) (unraveled paradox)	16	25	38	52	18	28
Interview Subgroup 1 (10)	6	10	6	10	5	4
Group 2 (40) (Saw no contradiction— claimed the barber shaved himself)	7	12	4	4	2	4
Interview Subgroup 2 (6)	3	2	1	3	2	3
Group 3 (28) (Gave up on the problem)	1	5	1	0	1	0
Interview subgroup 3 (4)	0	3	0	0	1	0

math, science, literature, languages, social studies etc. Even mathematics is increasingly viewed as a highly specialized field in spite of its intricate connections to the arts and sciences. The thinkers of the Renaissance did not view themselves simply as mathematicians, or inventors or painters, or philosophers or political theorists, but thought of themselves as seekers of Knowledge, Truth and Beauty. In other words there was a Gestalt world-view with polymaths that worked back and forth between multiple domains. The results of this three year study with 120 pre-service students indicate that nearly half of the students displayed polymathic traits—as a result of their attempt to unravel the given paradox. This suggests that interdisciplinary activities can certainly play an important role in the education of these future teachers. By taking a phenomenological approach and trying to understand the first person perspective of these students, deep set hidden beliefs about the nature of mathematics also became apparent through the voices of the students. The pre-service teachers also reported an increased interest in the place of paradoxes in mathematics, which they had believed as an infallible or absolutist science.

However the significance and applicability of these findings come with certain limitations. It should be noted that nearly one fourth of the students (group 3) in the study experienced and reported negative affective experiences as a result of tackling the paradox, which need to be sensitively attended to by the teacher educator. Debriefing sessions conducted during the course of the study were essential to create a positive pedagogical atmosphere on these students and foster a willingness on their part to try interdisciplinary activities in their classroom which integrates mathematics with other subjects. Another missing ingredient in this study voiced by many students in their journal writings and interviews was the need to engage in discourse with other students when confronted with the paradox. It would be of interest to the community of researchers to investigate how pre-service elementary teachers tackle paradoxes in a collaborative group effort. Due to the limitations in the resources available for this study, the author did not pursue this approach but this remains a fertile area for further investigation. Finally, many of these students voiced concern over concrete ways in which interdisciplinary activities could be introduced in the elementary classroom. To this end there have been recent attempts to classify works of mathematics fiction suitable for use by K–12 teachers in conjunction with science and humanities teachers to broaden student learning. Padula (2005) argues that although good elementary teachers have historically known the value of mathematical fiction, mainly picture books, through which children could be engaged in mathematical learning, such an approach also has considerable value at the secondary level. Padula (2005) provides a small classification of books appropriate for use at the middle and high school levels, which integrate paradoxes, art, history, literature and science to "stimulate the interest of reluctant mathematics learners, reinforce the motivation of the student who is already intrigued by mathematics, introduce topics, supply interesting applications, and provide mathematical ideas in a literary and at times, highly visual context" (p. 13)

Mathematical paradoxes played a significant role in the historical development of the field. These paradoxes contain enormous pedagogical potential for pre-service teacher education, particularly in showing that even mathematics can be fallible. As seen in this study, realizing the fallibility within what students believed was an absolutistic and monolithic structure, these prospective teachers experienced both a sense of empowerment and expressed changing views about the nature of mathematics. It is the authors hope that this line of research is further developed by the community of researchers and teacher educators to make a positive impact on the beliefs and practices of future teachers of mathematics.

REFERENCES

Annells, M. (2006). Triangulation of qualitative approaches: hermeneutical phenomenology and grounded theory. *Journal of Advanced Nursing, 56*(1), 55–61.

Australian Education Council (1990). *A national statement on mathematics for Australian schools.* Melbourne, VC: Australian Educational Council.

Ball, D. L. (1990). The mathematical understandings that pre-service teachers bring to teacher education. *Elementary School Journal, 90,* 449–466.

Cobb, P. (1988). The tension between theories of learning and theories of instruction in mathematics education. *Educational Psychologist, 23,* 87–104.

Ernest, P. (1989). The impact of beliefs on the teaching of mathematics. In C. Keitel, P. Damerow, A. Bishop & P. Gerdes (Eds.), *Mathematics, Education, and Society* (pp. 99–101). Paris: UNESCO Science and Technology Education Document Series No 35.

Ernest, P. (1991). *The philosophy of mathematics education.* London: Falmer Press.

Fenstermacher, G.D. (1978). A philosophical consideration of recent research on teacher effectiveness. In L. S. Shulman (Ed.), *Review of research in education 6* (pp. 157–185). Ithasca, IL: Peacock.

Goldin, G. A. 2000. Affective pathways and representations in mathematical problem solving. *Mathematical Thinking and Learning, 17*(2), pp. 209–219.

Goldin, G. A. (2002). Affect, meta-affect, and mathematical belief structures. In G. Leder, E. Pehkonen & G. Törner (Eds.), *Beliefs: A hidden variable in mathematics education?* (pp. 59–72). Dordrecht: Kluwer Academic Publishers.

Grigutsch, S. (1996). "Mathematische Weltbilder" bei Schülern: Struktur, Entwicklung, Einflussfaktoren. Dissertation. Duisburg: Gerhard-Mercator-Universität Duisburg, Fachbereich Mathematik.

Leathman, K. (2006). Viewing mathematics teacher beliefs as sensible systems. *Journal of Mathematics Teacher Education, 9*(1), 91–102.

Leder G. C., Pehkonen E., & Törner G. (Eds.), (2002). *Beliefs: A hidden variable in mathematics education?* (Vol. 31). Dodrecht: Kluwer Academic Publishers.

Merleau-Ponty, M. (1962). *Phenomenology of perception* (C. Smith, Trans.). London: Routledge & Kegan Paul.

National Council of Teachers of Mathematics (2000). *Principles and standards for school mathematics.* Reston, VA: Author.

Padula, J. (2005). Mathematical fiction—It's place in secondary school mathematics learning. *The Australian Mathematics Teacher, 61*(4), 6–13.

Pajares, M.F. (1992). Teachers' beliefs and educational research: Cleaning up a messy construct. *Review of Educational Research, 62*(3), 307–332

Root-Bernstein, R. S. (1989). *Discovering.* Cambridge, MA: Harvard University Press.

Root-Bernstein, R. S. (1996). The sciences and arts share a common creative aesthetic. In A. I. Tauber (Ed.), *The elusive synthesis: Aesthetics and science* (pp. 49–82). Netherlands: Kluwer.

Root-Bernstein, R. S. (2000). Art advances science. *Nature, 407,* 134.

Root-Bernstein, R. S. (2001). Music, science, and creativity. *Leonardo, 34,* 63–68.

Root-Bernstein, R. S. (2003). The art of innovation: Polymaths and the universality of the creative process. In L. Shavinina (Ed.), *International handbook of innovation* (pp. 267–278), Amsterdam: Elsevier.

Sriraman, B. (2003a). Can mathematical discovery fill the existential void? The use of Conjecture, Proof and Refutation in a high school classroom, *Mathematics in School, 32*(2), 2–6.

Sriraman, B. (2003b). Mathematics and literature: Synonyms, antonyms or the perfect amalgam. *The Australian Mathematics Teacher, 59*(4), 26–31.

Sriraman, B. (2004). Mathematics and literature (the sequel): Imagination as a pathway to advanced mathematical ideas and philosophy. *The Australian Mathematics Teacher. 60*(1), 17–23.

Sriraman, B. (2005). Re-creating the Renaissance. In M. Anaya, & C. Michelsen (Eds.), *Relations between mathematics and others subjects of art and science*. Proceedings of the 10th International Congress of Mathematics Education, Copenhagen, Denmark (pp. 14–19).

Sriraman, B., & Dahl, B. (2007). On bringing interdisciplinary ideas to gifted education. In press in L.V. Shavinina (Ed.), *The international handbook of giftedness*. Springer Science.

Thompson, A. (1992). Teachers' beliefs and conceptions: A synthesis of the research. In D.A.Grouws (Ed). *Handbook of research on mathematics teaching and learning* (pp. 127–146). Simon & Schuster and Prentice Hall International.

Törner, G. (2002). Mathematical beliefs. In G. C. Leder., E. Pehkonen, E., & G. Törner (Eds.). *Beliefs: A hidden variable in mathematics education?* (pp. 73–94). Dordrecht: Kluwer Academic Publishers.

Romme, M. A. J., & Escher, A. D. M. A. C. (1993). The new approach: A Dutch experiment. In M. A. J. Romme & A. D. M. A. C. Escher (Eds.), *Accepting voices* (pp. 11–27). London: MIND Publications.

Sternberg, R. J., Grigorenko, E. L., & Singer, J. L. (Eds.) (2004). *Creativity: From potential to realization*. Washington, DC: American Psychological Association.

Wedege, T., & Skott, J. (2006). *Changing views and practices: A study of the KappAbel mathematics competition*. Research Report: Norwegian Center for Mathematics Education & Norwegian University of Science and Technology, 274 pp. Trondheim

ACKNOWLEDGMENT

Reprint of Sriraman, B. (2007). Paradoxes as pathways into polymathy and discovery of mathematical structures. In B. Sriraman, C. Michelsen et al. (Eds.) *Proceedings of the 2nd International Symposium on Mathematics and is Connections to the Arts and Sciences*, Odense, Denmark. Reprinted with permission from Bharath Sriraman.

SECTION III

GEOMETRY AND HISTORY

CHAPTER 8

VORONOI DIAGRAMS

Michael Mumm

STATEMENT OF THE PROBLEM

Suppose we have a finite number of distinct points in the plane. We refer to these points as sites. We wish to partition the plane into disjoint regions called cells, each of which contains exactly one site, so that all other points within a cell are closer to that cell's site than to any other site.

An example of a Voronoi diagram:

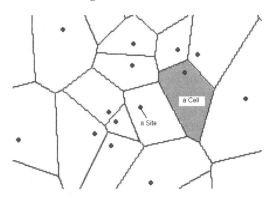

Interdisciplinarity, Creativity, and Learning, pages 97–108

Stated more formally, suppose P: = {p$_1$, p$_2$, . . . , p$_n$} is a set of distinct points (sites) in the plane. We subdivide the plane into n cells so that each cell contains exactly one site. An arbitrary point (x,y) is in a cell corresponding to a site p$_i$ with coordinates (x_{pi}, y_{pi}) if and only if

$$\sqrt{(x - x_{pi})^2 + (y - y_{pi})^2} < \sqrt{(x - x_{pj})^2 + (y - y_{pj})^2}$$

for all p$_j$ with $j \neq i$, $1 \leq j$, $i \leq n$. That is, the Euclidean distance from (x,y) to any other site is greater than the distance from (x,y) to p$_i$.

It turns out that the boundaries of the cells defined in this way will be composed of straight lines and segments forming convex polygons and will be defined by the perpendicular bisectors of segments joining each pair of sites. This method of partitioning a plane is called a *Voronoi diagram.*

Although this paper deals chiefly with two dimensional diagrams and the Euclidean distance metric, it should be noted that the concept of Voronoi diagrams can be generalized to n dimensions and to an arbitrarily defined distance metric. In addition, general geometric primitives such as line segments or curves may be used as sites instead of ordinary one-dimensional points. In the course of our discussion we will describe techniques of constructing Voronoi diagrams, provide some historical background on the subject, and discuss the multitude of applications that utilize Voronoi diagrams.

RATIONALE

Voronoi diagrams have countless applications in nearly all of the major sciences. Whenever one has a discrete set of data distributed in such a way that the concept of 'distance' has some meaning, a Voronoi diagram may be useful. With a Voronoi diagram as a reference, it is unnecessary to calculate the distance to each site in order to determine which site is closest to a particular point. The site corresponding to the cell that contains the point will always be closest. In applications which have a fixed set of sites, a Voronoi diagram need only be constructed once and then all subsequent distance calculations become unnecessary. Even if more sites are eventually added to a system, the basic structure of the Voronoi diagram remains intact. It is relatively easy to modify an existing diagram to accommodate new sites without reconstructing the entire system. For large scientific projects which use computers, this reduction in basic operations may result in a dramatic increase of algorithmic efficiency.

Voronoi diagrams may be extremely useful in the business world as well. A typical example is a pizza delivery franchise which has a network of restaurants servicing a large city. When an order comes into the central office, the operator, or computer, can use a Voronoi diagram to determine which restaurant will be able to deliver the pizza quickest relative to the location of the caller. It's easy to imagine similar applications arising in a large mail-order company with many distribution warehouses like amazon.com, or in the postal service itself.

We continue the discussion of applications throughout the paper.

HISTORY/BACKGROUND

History

Voronoi diagrams have a long history, dating back as early as the 17th century. Work by Descartes on a partitioning of the universe into "vortices" is one of the first known references to the subject. Even though Descartes does not explicitly define his vortices in the same way as Voronoi cells, his work is conceptually very similar [3] (Figure 8.2).

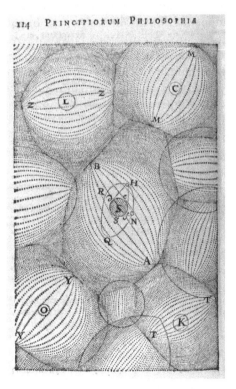

Figure 8.2 A drawing from Descartes which describes the partitioning of the universe into vortices—Notice that the vortices closely resemble Voronoi cells[11]:

Two German mathematicians, Lejeune Dirichlet and M. G. Voronoi, were credited with formalizing the modern concept of the Voronoi diagram [5]. Dirichlet was born in 1805 and in his work on quadratic forms he made some of the first significant contributions to the field of Voronoi diagrams. Indeed, it is because of him that Voronoi diagrams are also well known as Dirichlet tessellations. Before his death in 1859, Dirichlet had formalized the concept of the Voronoi diagram in the two and three dimensional cases [5]. Work by M. G. Voronoi in 1908 formalized the n dimensional case and gave Voronoi diagrams the name we commonly use today.

The two dimensional dual of the Voronoi diagram in a graph theoretical sense is the Delaunay triangulation [2]. Work on Delaunay triangulations (or, alternately, Delaunay tessellations) was done by French mathematician Charles Delaunay before 1872. In a Delaunay triangulation, any two sites are connected if they share a Voronoi diagram cell boundary, as shown in Figure 8.3. An alternate definition, more in accordance with Delaunay's original work, is that two sites are connected if and only if they lie on a circle whose interior contains no other sites [3].

Even before Voronoi diagrams were formalized mathematically, they were developed independently in other sciences. In 1909, BT Boldyrev a Russian scientist, used "area of influence polygons" in his work in Geology [10]. Voronoi diagrams were used in Meteorology by Thiessen in 1911 to help model average rainfall [5]. Influential work in crystallography was done utilizing Voronoi diagrams by a German named Paul Niggli in 1927. In 1933, physicists EP Wigner and F. Seitz did important research using Voronoi diagrams in physics. Voronoi diagrams continued to play a key role in research done in Physics, Ecology, Anatomy, and Astronomy throughout the 1900s.

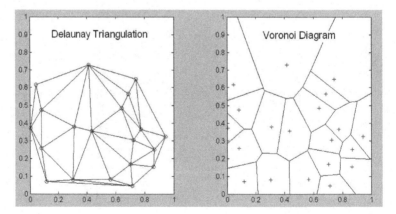

Figure 8.3 An example of a Delaunay triangulation with its corresponding Voronoi Diagram.

Applications and Algorithms

As mentioned earlier, applications of Voronoi diagrams are by no means confined strictly to mathematics. They go by many names as they relate to various fields of science. We provide a small glossary below [2,3,16].

Field of science:	Term used:
Mathematics:	Voronoi diagram, Dirichlet tessellation
Biology and Physiology:	Plant polygons, Capillary domains, Medial axis transform
Chemistry and Physics:	Wigner-Seitz zones
Crystallography	Domains of action, Wirkungsbereich
Meteorology and Geography:	Thiessen polygons

In the Work/Solutions section of this paper, we describe a way of constructing a Voronoi diagram using a geometric algorithm. It turns out that although this kind of algorithm is very easy to intuitively understand, it is not as computationally efficient as other known techniques. In [3], Aurenhammer and Klein classify some of these algorithms. The geometric technique we describe is an example of an *incremental construction* algorithm and it has a relatively poor algorithmic efficiency of $O(n^2)$. Using this notation, we assume the reader has a basic understanding of algorithmic efficiency classification. In these examples, n is the number of sites in the system.

Another category of algorithm used for Voronoi diagrams is *divide and conquer*. This technique works by recursively dividing the set of sites in order to decrease the problem size. Eventually the subsets of sites are small enough that diagrams are easily constructible. These sub-diagrams then must be merged back together, up the recursive tree, into the complete diagram for the system. Although this merging process is complicated, and care needs to be given as to how the set of sites is split each time, the result is a total algorithmic efficiency of $O(n*log(n))$. This is a significant improvement over the incremental construction algorithm.

Another algorithm of $O(n*log(n))$ efficiency is the *sweep* algorithm. This algorithm works by sweeping a vertical line across the plane horizontally. As this line passes through various Voronoi cells, their boundaries are constructed. Although the efficiency of this algorithm is the same as the *divide and conquer*, it can have some additional desirable characteristics if the sites are distributed in a certain way. There is no universally superior algorithm; ultimately one must choose an algorithm based on the particulars of the data and of the application.

WORK/SOLUTIONS

Our first task is to describe a simple geometric algorithm for constructing a Voronoi diagram. We start with a system containing only two sites and build to the four site case. From here it is easy to generalize to any system with a finite number of sites.

A Simple Geometric Construction Algorithm

As mentioned in the first section, perpendicular bisectors have a fundamental role in the construction of a Voronoi diagram. When our system contains only two sites it is a very easy matter to construct a Voronoi diagram; the perpendicular bisector of the segment joining the two sites is all that is necessary (Figure 8.4b).

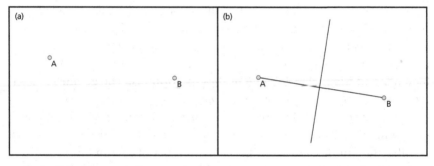

To build a Voronoi diagram for three sites (points A, B, D below) we first construct a triangle with vertices at the three sites. From here we must rely on the fact that the perpendicular bisectors of the three sides of a triangle meet at a single point. This is illustrated in Figure 8.4c. To complete the diagram we simply remove the superfluous rays, line segments, and points which were used in construction (Figure 8.4d).

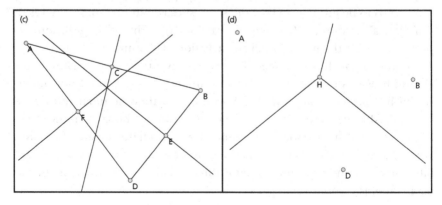

Things become a bit more complicated when we have four sites. We begin with the existing diagram for three sites, then add another site (point I below) and adjust the diagram accordingly. To do this we must construct perpendicular bisectors between I and each previously existing site. These bisectors allow us to determine which parts of the diagram need to be adjusted.

We begin by constructing the perpendicular bisector between I and D (Figure 8.4e). From this, we see that everything to the left of the line KJ belongs to I's cell. We adjust the diagram accordingly and remove the superfluous parts (Figure 8.4f).

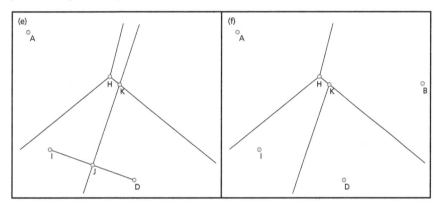

Now we construct the bisector between I and A (Figure 8.4g). From this we see that everything below the line LN belongs to I's cell. Now we remove the superfluous parts (Figure 8.4h).

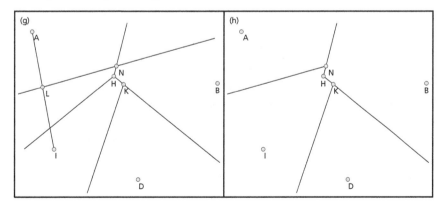

Finally, we construct the bisector between I and B, our last pair of sites (Figure 8.4i). From this we see that everything to the left of segment NK belong to I's cell. After removing the remaining superfluous parts, we have our completed Voronoi diagram for four sites (Figure 8.4j).

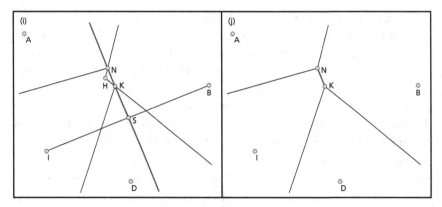

This kind of approach is sufficient for any number of sites. From the four site system above we can construct a five site system by adding a site and considering which boundaries need to be adjusted for each new pair of sites. Continuing in this way we can build our system as large as we wish.

An Example Problem

Construct a Voronoi diagram using the vertices of a regular pentagon as sites (Figure 8.5).

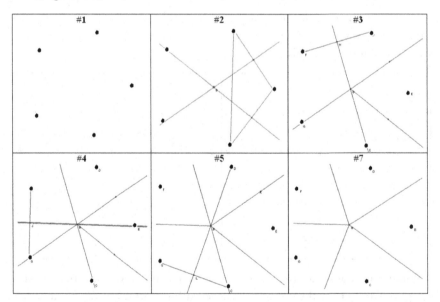

Figure 8.5

Voronoi of the Missoula Area

I decided it would be interesting to construct a Voronoi diagram of sites with local geographical relevance. After a good deal of searching, I was able to track down some data through the Montana Natural Resource Information System's GIS web service [7]. Using their service I was able to produce a list of the longitude and latitude of some 236 cities and towns in the Missoula, Montana area. This data then needed to be formatted so that it could be plotted in a Euclidean space. After that, MATLAB's Voronoi package was used to produce the actual diagram (Figure 8.6). Finally, I labeled a few key communities based on size of population and range of disbursement, so that one could get a good geographic impression of how the data was distributed. The result would have been too cluttered if I had labeled every site.

One thing the diagram provides is a sense of how remote certain sites are. If a city or town has few neighboring communities, its Voronoi cell is generally large. For example, Seeley Lake has a large Voronoi region corresponding to its relative remoteness. Even though the diagram gives us no true indication of population, we can make a rough extrapolation based on the fact that, in general, urban areas of high population densities have nu-

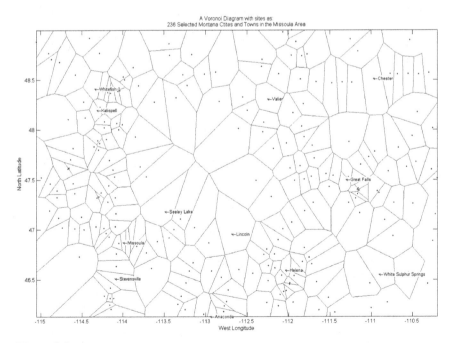

Figure 8.6

merous nearby neighbors. Thus, we would expect larger cities to occur in areas of the diagram with smaller cells. This is more or less consistent with what the diagram indicates.

A more practical application of a diagram like this might be related to school districting. If each site on our diagram corresponded to a community with a school, the boundaries of the cells would be effective districting lines. If I child lived in a particular cell, they would go to school in the community that corresponded to that cell because it would be the shortest distance away.

CONCLUSIONS AND IMPLICATIONS

As we have seen, Voronoi diagrams have been around a long time and have undergone a good deal of study. In fact, Aurenhammer and Klein claim that about one out of 16 papers in computational geometry have been on research concerning Voronoi diagrams! More than 600 papers on the subject are listed in [8]. Ever since Descartes partitioned the universe into vortices in 1644, research in Voronoi diagrams has been done by some of the brightest minds in science and mathematics.

Because of the importance of Voronoi diagrams, the efficiency of computer algorithms used in diagram construction is equally important. There has been a good deal of progress made in algorithm design already and popular modern algorithms are extremely efficient. Even so, there is room for improvement by adapting algorithms to take advantage of known characteristics (about distribution etc.) in a set of data or a type of application.

We have hinted at a number of general applications of the Voronoi technique, but it should be made clear that there are vastly more that have gone without mention. Because of their generality, Voronoi diagrams are useful in an extremely diverse array of situations, in many different aspects of science and of everyday life. Here are a just few other applications, listed in [9]:

- Voronoi diagrams may be used in computer GUI applications that need to determine which link is closest to the mouse cursor when a user clicks on the screen.
- Voronoi diagrams have been used to correspond distribution of Bark Beetle attacks on trees to the beetle's known territorial behavior. The beetles tend to feed in isolated Voronoi regions, in a way that discourages intra-species competition for the same resources [12].
- Voronoi diagrams are used in some graphics software to create "Non-Photorealistic" images. The software distorts the original image around Voronoi cells of selected points.

- Voronoi diagrams have been investigated as a means of grouping words and multi-part symbols in documents [13].
- Voronoi diagrams have been used to increase the accuracy in the "river-mile" positioning system often employed along rivers and in reservoirs. Given an arbitrary position on the water, Thiessen polygons are used to determine which known river mile position is closest [14].
- Voronoi diagrams of a collection of randomly distributed points are being used to model some kinds of steel [15].

It is possible to generalize Voronoi diagrams to three or more dimensions and to define our distance metric however we wish. For instance, we could use a taxicab distance metric to produce a more useful diagram of sites in a city. Since it is seldom possible to travel directly to a destination in a city, a Voronoi diagram using the taxicab metric would more accurately reflect the amount of time it takes to travel along city streets to a particular site. In addition, we could weigh the 'distance' of certain busy streets more than streets in other less trafficked areas so that our diagram would even more accurately reflect travel time. The exceptional adaptability of the Voronoi diagram concept is what makes it so versatile and widely used throughout the sciences.

WORKS CITED

1. Index of Biographies. 12 Nov. 2003. School of Mathematics and Statistics, University of St. Andrews, 1996 <http://www-gap.dcs.st-and.ac.uk/~history/BiogIndex.html>
2. Voronoi Diagram. 12 Nov. 2003. Wolfram Research, 2003. <http://mathworld.wolfram.com/VoronoiDiagram.html>
3. Aurenhammer, Franz and Klein, Rolf. "Voronoi Diagrams." <wwwpi6.fernuni-hagen.de/publ/tr198.pdf> FernUniversität in Hagen
4. Meyer, Walter. Geometry and Its Applications. New York: Harcourt Academic Press, 1999
5. Voronoi Diagrams in Biology. 15 Nov. 2003. Beloit College, 5/15/98 <http://biology.beloit.edu/students/zdravko/vor_history.html>
6. Voronoi Diagrams: The Post Office Problem. 15 Nov. 2003. <www.comp.lancs.ac.uk/~kristof/research/ notes/voronoi/voronoi.pdf>
7. The Montana Natural Resource Information System Geographic Information System (GIS). 22 Nov. 2003. Montana Natural Resource Information System, 2003, <http://nris.state.mt.us/gis/default2.htm>
8. A. Okabe, B. Boots, and K. Sugihara. Spatial Tessellations: Concepts and Applications of Voronoi Diagrams. John Wiley & Sons, Chichester, UK, 1992.
9. Geometry in Action. 22 Nov. 2003. David Eppstein, ICS, UC Irvine, 2003, <http://www.ics.uci.edu/~eppstein/gina/voronoi.html>

10. Fast Generalizations of Voronoi Diagrams Using Graphics Hardware. 23 Nov. 2003. Kenneth E. Hoff III, University of North Carolina. 10/6/99 <http://www.cs.unc.edu/~geom/voronoi/>

11. Heavens (World Treasures of the Library of Congress: Beginnings). 12 Mar. 2004. <http://www.loc.gov/exhibits/world/heavens.html>

12. Dirichlet tessellation of bark beetle spatial attack points. 15 Mar. 2004. Byers, J.A., 1992. <http://www.wcrl.ars.usda.gov/cec/dir-abs.htm>

13. Using the Voronoi Tessellation for Grouping Words and Multi-part Symbols in Documents. 12 Mar. 2004. M. Burge and G. Monagan, Department of Computer Science, Swiss Federal Institute of Technology. 3 Jul. 1997 <http://www.computing.armstrong.edu/FacNStaff/burge/publications/papers/visgeo–95/>

14. Dynamic Segmentation and Thiessen Polygons: A Solution to the River Mile Problem. 17 Mar. 2004. William W. Hargrove, Richard F. Winterfield, Daniel A. Levine, Office of Environmental Restoration and Waste Management. <http://research.esd.ornl.gov/CRERP/DOCS/RIVERMI/P114.HTM>

15. Modeling of Microstructures by Voronoi cells. 17 Mar. 2004. Mikael Nygårds and Peter Gudmundson. Department of Solid Mechanics KTH, Stockholm, Sweden. <http://www.brinell.kth.se/CONF99/Mikael_Nygards.html>

16. Voronoi Diagrams: Applications from Archaology to Zoology. 18 Mar. 2004. Scot Drysdale, Dartmouth College. <http://www.ics.uci.edu/~eppstein/gina/scot.drysdale.html>

ACKNOWLEDGEMENT

Reprint of Mumm, M. (2004). Voronoi Diagrams. *The Montana Mathematics Enthusiast*, *1*(2), 44–55. © The Montana Mathematics Enthusiast.

This paper is the developed version of a term paper written in Euclidean & Non Euclidean Geometries, taught by Professor Bharath Sriraman at The University of Montana in Fall 2003.

CHAPTER 9

AN IN-DEPTH INVESTIGATION OF THE DIVINE RATIO

Birch Fett
The University of Montana

ABSTRACT

The interesting thing about mathematical concepts is that we can trace their development or discoveries throughout history. Most cultures of the ancient world had some form of mathematics, and these basic skills developed into what we now call modern mathematics. The divine ratio is similar in that it was used in many different sections of history. The divine ratio, sometimes called the golden ratio or golden section, has been found in very diverse areas. The mathematical concepts of the golden ration have been found throughout nature, in architecture, music as well as in art. Phi is an astonishing number because it has inspired thinkers in many disciplines, more-so than any other number has in the history of mathematics. This paper investigates how the golden ratio has influenced civilizations throughout history and has intrigued mathematicians and others by its prevalence.

INTRODUCTION

Throughout this paper, the terms golden ratio, divine ratio, golden mean, golden section and Phi (ϕ) are interchangeably used. Wasler, (2001) de-

Interdisciplinarity, Creativity, and Learning, pages 109–131
Copyright © 2009 by Information Age Publishing
109

fines the golden ratio as a line segment that is divided into the ratio of the larger segment being related to the smaller segment exactly as the whole segment is related to the larger segment. The divine ratio is the ratio of the larger segment, AB, of line AC to the smaller segment BC of the line AC.

This same definition was first given by Euclid of Alexandria around 300 BC. He defined this proportion and called it "extreme and mean ratio" (Livio, 2002). Let us assume that the total length of line AC is $x + 1$ units and the larger segment AB has a length of x. This would mean that the shorter segment BC would have a length of 1 unit. Now we can set up a proportion of AC/AB = AB/BC.

$$\frac{x+1}{x} = \frac{x}{1}$$

By cross multiplying it yields $x^2 - x - 1$. Using the quadratic formula, two solutions become apparent $(1 + \sqrt{5})/2$ and $(1 - \sqrt{5})/2$, and we only use the positive solution because we are in terms of a length. The positive solution is $(1 + \sqrt{5})/2$. Phi is the only number that has the unique property that $\phi * \phi' = -1$ where ϕ' is the negative solution to the quadratic $(1 - \sqrt{5})/2$ (Huntley, 1970).

ADDITIONAL INFORMATION ON THE GOLDEN RATIO

In professional mathematical literature, the golden ratio is represented by the Greek letter tau. The symbol (τ) means "the cut" or "the section" in Greek. In the early twentieth century, American mathematician gave the golden ratio a new name. Mark Barr represented the Golden Ratio as phi (ϕ), which is the first Greek letter in the name of Phidias.[1] Barr chose to honor the great sculptor because many of Phidias's sculptors contained the Divine Ratio.

The golden ratio is a known irrational number. Irrational numbers have been around for sometime. Most historians believe that irrational numbers were discovered in the fifth century BC. Pythagoreans knew about irrational numbers and believed that the existence of such numbers was due to a cosmic error (Livio, 2001).

The Golden section is aesthetically pleasing in nature. Phi represents some remarkable relationships between the proportions of patterns of living plants and animals. Contour spirals of shells, such as the chambered nautilus, reveal growth patterns that are related to the golden ratio. The nautilus shell has patterns that are logarithmic spirals[2] of the golden section. Each section is characterized by a spiral, and the new spiral is extremely close to the proportion of the golden section square larger than the previous. The growth patterns in nature approach the golden ration, and in some cases come very close to it, but never actually reach the exact proportion (Elam, 2001). A construction of the golden rectangle and logarithmic spiral can be seen in Figure 9.2.

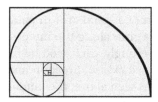

Figure 9.2

Logarithmic spirals can be found through-out nature. Ram horns and elephant tusks, although they do not lie in a plane, follow logarithmic spirals. Logarithmic spirals are also closely related to Golden Triangles.[3] Starting with a Golden Triangle ABC, the bisector of angle B meets AC at point D and is the golden cut of AC. With this bisection, triangle ABC has been cut into two isosceles triangles that have golden proportion (the ratio of their areas is ϕ:1. Continuing this process by bisecting angle C, point E is obtained. Again point E is the golden cut along line BD, thus constructing two more golden triangles. This process produces a series of gnomons[4] that will eventually converge to a limiting point O, which is the pole of a logarithmic spiral passing successively and in the same order through the three vertices (. . . A, B, C, D . . .) of each of the series of the triangles (Huntley, 1970).

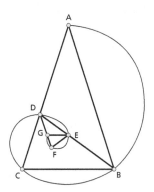

Figure 9.3

If we begin with GF and call it the unit length, then:

FE = 1ϕ
ED = $1\phi + 1$
DC = $2\phi + 1$
CB = $3\phi + 2$
BA = $5\phi + 3$

By bisecting the base angles of the successive gnomons, the lengths of these segments form a Fibonacci series, which we have already seen to converge to the Golden Ratio.

Pine cones and sunflowers are closely related to the Golden Ratio. Each seed in a pine cone is part of a spiral growth pattern that closely relates to ϕ. The seeds of pine cones grow along two intersecting spirals that move in opposite directions. Interestingly, each seed belongs to both spirals. Eight of the spirals move in the clockwise direction and the remaining thirteen move counter clockwise. As seen above, the numbers 8 and 13 are consecutive Fibonacci numbers which converge to the Golden Ratio. The proportion of 8:13 is 1:1.625. Sunflowers exhibit the same spiral patterns as seen in pine cones. Sunflowers have 21 clockwise spirals and 34 counter clockwise. The proportion of 21:34 is even closer to the Golden Ratio than that of pine cones; it is 1:1.619 (Elam, 2001).

The geometry of plant axis flexure is the result of orthotropic growth and the stress caused by a vertical weight distribution along the axis. A flexed plant axis is shown to conform to a portion of a logarithmic spiral. With numerous plants, this mode of curvature is the most prevalent condition of plants lacking or having secondary growth. Plants like sunflowers represent this growth pattern (Niklas and O'Rourke, 1982).

In 1907 the German mathematician G. van Iterson showed that the human eye would pick out patterns of winding spirals when successive points were packed tightly together. The points were separated by the Golden Angle which measures to 137.5 degrees. The familiar spirals that the human eye would pick out consisted of counter clockwise and clockwise patterns of consecutive Fibonacci numbers. Nature, specifically sunflowers, grows in the most efficient way[5] of sharing horizontal space, which is in proportion of the Golden Ratio. Most sunflowers have a 21:34 ratio, but few have been reported with proportions of 89:55, 144:89 and 233:144 (Livio, 2001).

The Golden Ratio can be found in many examples throughout the world. Phi can be seen in many places; from the layout of seeds in an apple to Salvador Dali's painting "Sacrament of the Last Supper" (Livio, 2002). In the following sections, an in-depth look is taken on the occurrences of Phi in as well as the development of Phi throughout history.

THE GOLDEN RATIO AND FIBONACCI NUMBERS

Leonardo de Pisa, born around 1175 AD, commonly known as Fibonacci[6] introduced the world to the rabbit problem. The rabbit problem asked to find the number of rabbits after n months, given that adult rabbits produce a pair of rabbits each month, offspring take one month to reach reproductive maturity, and that all the rabbits are immortal. This problem gave the mathematical world the series of Fibonacci numbers.[7]

In 1202 AD Fibonacci wrote, *Liber Abaci*, which was a book based on the arithmetic and algebra that he had accumulated in his travels. This book was widely copied and introduced the Hindu-Arabic place-value decimal system and the use of Arabic numerals into Europe. Most of the problems in *Liber Abaci* were aimed at merchants and related to the price of goods, how to calculate profit on transactions, and how to convert between the various currencies in use in the Mediterranean countries. Fibonacci is most remembered for presenting the world with the "rabbit problem" which is located in the third section of *Liber Abaci*.

Looking at the ratio of successive Fibonacci numbers, an interesting value appears. As the n increases, the ratio of F_n/F_{n+1} approaches the golden ratio. The values ($n = 1 \ldots 10$) can be seen in the table below:

n	F(*n*)	F(*n*)/F(*n* − 1)
1	0	
2	1	
3	1	1
4	2	2
5	3	1.5
6	5	1.666667
7	8	1.6
8	13	1.625
9	21	1.615385
10	34	1.619048

The convergence of Fibonacci numbers to the Golden Ratio can be seen in the "rabbit problem." A Scottish mathematician, in the early 1700s made the connection between the Golden Ratio and the rabbit problem. Robert Simson (1687–1768), noticed that consecutive terms of the solution to the rabbit problem converged to the Golden Ratio (Johnson, 1999). A geometric sequence can be constructed on the basis of the breeding rabbits. For example, let adult rabbits be represented by 'A' and their offspring represented by 'b'. The arrangement of adults (A) and their offspring (b), can be written as AbAAbAbAAbAAbAbAAbAbA . . . The sequence of A's and b's may be extended indefinitely in a unique way because the rule for generating the next character is well defined. The ratio of adults to offspring rabbits in the limit of an infinite sequence is equal to the Golden Ratio (Dunlap, 1997).

$$\lim_{n} \to \phi \; A/b = \phi$$

THE GOLDEN RATIO IN ANCIENT GREECE

The Golden Ratio can be found throughout nature, which will be discussed below, but it can also be found in the history of the heavens. Plato (428–347 BC) prophesied the significance even before Euclid described it in *Elements*. Plato saw the world in terms of perfect geometric proportions and symmetry. His ideas were based on Platonic Solids.[8] He divided the heavens into four basic elements, earth, water, air, and fire. Each of these elements was assigned a Platonic Solid; a cube for earth, tetrahedron for fire, octahedron for air and an icosahedron for water. Using this foundation, Plato created a chemistry that is similar to modern day chemistry[9] (Livio, 2003).

The five Platonic solids are the only existing solids in which all of the faces are identical and equilateral and each vertex is convex. Interestingly, each of the solids can be circumscribed by a sphere with all of its vertices lying of the sphere. The tetrahedron consisted of four triangular faces, the cube with six square faces, the octahedron with eight triangular faces, the dodecahedron with twelve pentagonal faces and the icosahedron with twenty triangular faces (Livio, 2002).

Each face of the regular polyhedron is a regular polygon with n edges. It is known that the values of n are {n: $3 \leq n < \infty$} with n being related to the interior angle α.[10] Each vertex of the three dimensional polygon is defined by the intersection of a number of faces, m, where $m \geq 3$. In order for a convex vertex to be formed, $m\alpha < 360$ degrees.[11] There are only five combinations of integers that satisfy these equations and they correspond with the five Platonic Solids and are listed below[12] (Dunlap, 1997).

Solid	n	m	e	f	v
tetrahedron	3	3	6	4	4
cube (hexahedron)	4	3	12	6	8
octahedron	3	4	12	8	6
dodecahedron	5	3	30	12	20
icosahedron	3	5	30	20	12

The Golden Ratio is of relevance to the geometry of figures with five-fold symmetry. The dodecahedron and the icosahedron are of particular interest. If either one of these Platonic Solids are constructed with an edge length of one unit, it is easy to see the important role the Golden Ratio play in their dimensions (Dunlap, 1997).

Solid	Surface Area	Volume
dodecahedron	$15\phi/(3-\phi)$	$5\phi^3/(6-2\phi)$
icosahedron	$5\sqrt{3}$	$5\phi^5/6$

Plato and his foundations using Platonic Solids for the heavens may suggest that the Golden Ratio may have been known in ancient Greece. However, the full mathematical properties of Platonic Solids may not have been known in antiquity. Plato and his followers may have created and used Platonic Solids in the foundations of the universe based on sheer beauty.

Many authors researching ancient Greek mathematics are unsure if the works of Plato were influenced by Pythagoras and the Pythagoreans. Pythagoras[13] was born around 570 BC on the island of Samos. Pythagoras and the Pythagoreans are best known for their role in the development of mathematics and for the application of mathematics to the concept of order (Livio, 2002).

The Pythagoreans assigned special properties to odd and even numbers as well as individual numbers. The number one was considered the generator of all other numbers and geometrically, the generator of all dimensions. The number two was considered the first female number and the number of opinion and division. Geometrically, the number two was expressed by the line which has one dimension. The number three is considered by the Pythagoreans to be the first male number and the number of harmony because it combines the unity number (one) and the division number (two). The geometric expression of the number three was a triangle, where the area of the triangle has two dimensions. Justice and order was expressed in the number four. On the surface of the Earth, four directions provide orientation for humans to identify their coordinates in space. Four points, not in the same plane, form a tetrahedron. The number six is the first perfect number and considered the number of creation. It is the number of creation because it is the product of the first female number (two) and the first male number (three). Six is a perfect number because it is the sum of all the smaller numbers that divide into it. The first three perfect numbers are listed below (Livio, 2002).

$6 = 1 + 2 + 3$
$28 = 1 + 2 + 4 + 7 + 14$
$496 = 1 + 2 + 4 + 8 + 16 + 31$

The number five deserves its own explanation. Five represents the union of the first female number and the first male number. This union suggests that five is the number of love and marriage. The main reason five is important to this discussion is because the Pythagoreans used the pentagram[14] as a symbol of their brotherhood (Livio, 2002).

The construction of the pentagon, using a compass and marked straight edge, leads to a pentagram. Given a line AB, use the compass to draw arcs of radius a about points A and B. Next construct the perpendicular bisector PQ of line AB. Using the straight edge plot two points that are a units apart and slide the straight edge so that it passes through point A, until one of the points falls on the arc of B. There are only two possible positions for these points, namely, C and F. Using the same directions, find points G and D, sliding the straight edge through point B until one of the points falls on the arc of A. The fifth vertex (E) can be found by the requirement that on line EGB, EG equals *a*. Using this construction of a pentagon, one can connect the vertices and build a pentagram (Herz-Fischler, 1987) (Figure 9.4).

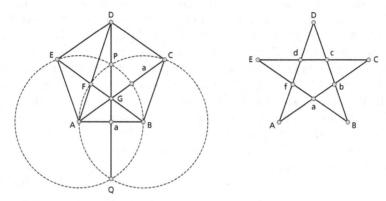

Figure 9.4

The pentagram is important to the discussion of the Golden Ratio because of its unique properties. The diagonals of a pentagon cut each other in the Golden Ratio and the larger of the two segments is equal to the side of the pentagon. The Pythagoreans choosing the pentagram as a symbol for brotherhood, and the given properties of the pentagram, suggests that the Pythagoreans were familiar with the Golden Number, but many historians are still under debate about this particular topic, due to inconclusive historical data (Herz-Fischler, 1987).

One theory, Heller (1958), suggests that the Pythagoreans used the pentagon to discover incommensurability and the division in extreme and mean ratio. Heller believes that the Pythagoreans discovered incommensurability through the observations of a series of pentagons when drawing diagonals.

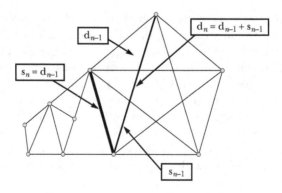

Figure 9.5

The diagonal, d_{n-1}, becomes the side, s_n, of the next largest pentagon. The new diagonal d_n is the sum of the side and the diagonal, s_{n-1} and d_{n-1}, of the previous pentagon. With this information it is easy to see the recurrence relationships $s_n = d_{n-1}$; $d_n = d_{n-1} + s_{n-1}$. Using[15] $s_1 = 2$ and $d_1 = 3$, leads to the sequence of $d_n : s_n$ ratios of $3/2, 5/3, 8/5, 13/8 \ldots$ which we have already seen to be successive Fibonacci numbers (Herz-Fischler, 1987). A formal proof of this can be found in *The Golden Ratio: The Story of Phi the World's Most Astonishing Number.*

THE GOLDEN MEAN IN ANCIENT EGYPT

Modern mathematicians have been trying to decide what civilizations used and understood the golden mean. Ancient Egypt, a civilization with profound mathematical accomplishments and astonishing monuments is under investigation for uses of the golden mean. Many interpretations of the golden mean use the properties of different geometrical figures. This may prove useless because it can produce an infinite chain of similar links. Math historians do need to focus on the ancient monuments and the mathematics of the respective time period. Ancient civilizations did not necessarily have the same numbering systems of modern times. This suggests that some things that work in modern numbering systems do not work in ancient systems (Rossi and Tout, 2002).

One theory about the use of the Golden Mean in ancient Egypt is that Egyptian architects designed the pyramids in a geometric way. Egyptian pyramids were based on geometrical processes of squares, rectangles and triangles. Of extreme importance was the process of the 8:5 triangles.[16] Egyptians used these triangles because the ratio of 8/5 was a good approximation of the Golden Mean. The theory continues to suggest that Egyptian architects gave their designs dimensions based on the corresponding numbers of the Fibonacci series. We have already seen that the ratio of corresponding Fibonacci numbers converges to the Golden Ratio (Rossi and Tout, 2002).

The Great Pyramid of Cheops, built before 2500 BC, has been measured and many different dimensions are present. The majority of the dimensions are within one percent of 755.79 feet as the length of the base and 481.4 feet as the height. Some theories claim that the Great Pyramid of Cheops was designed so that the ratio of the slant height of the pyramid to half the length of the base would be the divine proportion (Markowsky, 1992).

In Figure 9.6, h represents the height, b represents half the base, and s represents the slant height of the Great Pyramid of Cheops. Using 755.79 feet for the length of the base and 481.4 feet for the height, we can see that $b = 377.90$ feet. Using the Pythagorean Theorem, $h^2 + b^2 = s^2$, we can find that $s = 612.01$. This gives us a ratio of the slant height of the pyramid to

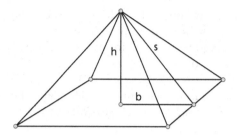

Figure 9.6

half the length of the base as $612.01/377.90 = 1.62$ which is very close to the Golden Mean (Markowsky, 1992). Another interesting feature of the Great Pyramid is that it has an apex angle of 63.43 degrees. This is very close to the apex angle of the Golden Rhombus[17] (63.435 degrees), which has dimensions derived for the Golden Ratio. The difference between the apex angle of the Great Pyramid and a Golden Rhombus is a mere 22 centimeters in the edge of the length of the pyramid base (Dunlap, 1997).

The question that needs to be answered is, was it possible for ancient Egyptians to construct a convergence of the Fibonacci numbers? Ancient Egyptians represented ratios as a sum of unit fractions. For example the fraction $3/5$ would be represented as $1/2 + 1/10$. As ratios continued to grow, many different representations become available. Take the ratio $13/21$ for example. Egyptians could have represented this number in five different ways:

1. $1/2 + 1/10 + 1/56 + 1/840$
2. $1/2 + 1/10 + 1/57 + 1/665$
3. $1/2 + 1/10 + 1/60 + 1/420$
4. $1/2 + 1/10 + 1/63 + 1/315$
5. $1/2 + 1/10 + 1/65 + 1/273$

Egyptian scribes could have found a convergence of ϕ with their system of representing fractions. Adding to the previous sum of ratios a unit fraction whose denominator is given by the multiplication of the two previous denominators (in the ratio of Fibonacci numbers) yields the next value in the sum converging to the Golden Ratio. The sum of the ratios of the first few Fibonacci numbers converging to ϕ can be seen below (Rossi and Tout, 2002).

$1/2 = 1/2$
$3/5 = 1/2 + 1/10$
$8/13 = 1/2 + 1/10 + 1/65$
$21/34 = 1/2 + 1/10 + 1/65 + 1/442$
$55/89 = 1/2 + 1/10 + 1/65 + 1/442 + 1/3026$
$144/233 = 1/2 + 1/10 + 1/65 + 1/442 + 1/3026 + 1/20737$

The convergence above suggests that is was possible for ancient Egyptian scribes to evaluate the Golden Ratio. However, it seems unlikely that ancient Egyptians were aware of the Fibonacci numbers. Egyptian math is considered an applied math, no records have been found on the theory behind their mathematics. Only applications of Egyptian mathematics exist. This suggests that the Egyptians, although capable, did not recognize the golden ratio and it was a mere coincidence that the architecture of the pyramids is based on 8:5 triangles (Rossi and Tout, 2002).

THE GOLDEN RATIO IN ANCIENT INDIA

The division in extreme and mean ratio appears in mathematical texts from India in connection with trigonometric functions. The Indian sine function is not the same as our modern day sine function. The Indian sine function can be defined as satisfying the relationship Sine $(\theta) = \frac{1}{2}*$ chord (2θ). The circumference of the circle is divided into 360 degrees and then the radius of the circle is divided into 60 parts. With this, sine $(30) = a_6/2 = r/2 = 30$. And sine $(18) = a_{10}/2$ and sine $(36) = a_5/2$.

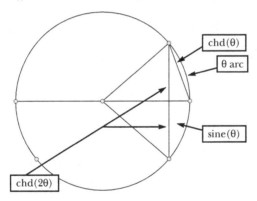

Figure 9.7

Bhaskara II (1114–1185) states without proof that Sine $(18) = (R(5r^2) - r)/4$. This is exactly the relationship sine $(18) = a_{10}/2$. Bhaskara, again without reason, tells to find the side of the pentagon inscribed in a circle, multiply the diameter by 70534/12000 (Amma, 1979). A proof of this statement is provided by Gupta (1976) and is provided below.

PROOF

In a circle of radius r = OX = OY, let the arc YM = 36 degrees. Draw a semicircle OX about the midpoint C of OX and draw the arc MD about Y. Assume that the tow arcs meet at the single point T on line YC.

Then Sine (18) = YM/2 = YT/2 = YC/2 − TC/2 = (R(r² + (r/2)²) − r/2)/2 which is equivalent to Sine (18) = (R(5r²) − r)/4.

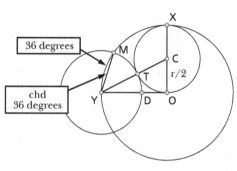

The above proof and construction are considered incomplete because they do not explain why the arcs meet at point T. Gupta (1976) continues and completes the construction by: Think of Y and C as given points and draw the arc OTX of radius r/2. Draw arc MTD of radius YT. Thus, the circles are tangent at the point T on the line YTC connecting the centers.

How does this construction tie in with the discussion on Ancient Indians knowing the Golden Ratio? Concentrate on the triangle YOC and arcs DT and OT. With a close examination, it can be seen that OY is divided in extreme and mean ratio at D. In other words, YM = YD is the greater segment when OY is divided in extreme and mean ratio (Gupta, 1976).

EVIDENCE OF THE GOLDEN RATIO IN THE ARTS

Countless illustrations of the proportions of the Golden Section are found in the works of humans. The Golden Section follows upon the basis of symmetry everywhere and the forms which are based upon the golden proportion are widely distributed. When speaking about the products of art and architecture, there is no equal symmetry, the artist or workmen unconsciously employ golden proportions. Irregular inequality and capricious division is aesthetically disagreeable, while golden proportions are pleasing to both hand an eye (Ackermann, 1895).

Many assertions claiming that the Golden Section was used in art are associated with the aesthetics of the proportion. When given an opportunity to choose the most visually pleasing rectangle, most people would choose rectangles with a close approximation of the Golden Rectangle. Although most humans cannot decipher between a rectangle with a ratio of 1.6 and a rectangle with ratio of 1.7, it suggests that humans do prefer rectangles in the range close to the Golden Rectangle (Markowsky, 1992).

Several decades after the Brotherhood of the Pythagoreans faded, the Golden Ratio continued to influence many artists and artisans. The Golden Ratio has influenced classical Greek architecture, notably the Parthenon in Athens. Inside the Parthenon stands a forty-foot-tall statue of the Greek Goddess Athena,[18] which has also shown to have Golden proportions. Both the temple and the stature were designed by Phidias, who is the first artist known to use the Golden Ratio in his work. As said above, the symbol for the Golden Ratio is the first Greek letter phi, which also happens to be the first letter in Phidias's name (Johnson, 1999).

Ancient Greek scholar and architect Marcus Vitruvius Pollio, who is commonly known as Vitruvius, advised that "the architecture of temples should be based on the likeness of the perfectly proportioned human body where a harmony exists among all parts" (Elam, 2001).Vitruvius is credited with introducing the concept of a module to the architectural world. This concept was the same as the module of human proportions and became an important architectural idea. The Parthenon[19] in Athens is an example of this proportioning. The Parthenon can be inscribed by a Golden Rectangle (Elam, 2001).

When the triangular pediment was still intact,[20] the Parthenon fit precisely into a Golden Rectangle. Another claim is that the height of the structure (from the top of the tympanum to the bottom of the pedestal) is divided into the Golden Ratio (Livio, 2001).Markowsky, (1992) has a contrasting view of the Parthenon. He believes that even though the Parthenon incorporates many geometric balances, its builders had no knowledge of the Golden Ratio. Depending on what sources are used, the dimensions of the Parthenon vary because the authors are measuring between different points. This implies that if the author is a Golden Ratio enthusiast they could choose which ever numbers give them the best approximation of ϕ.

Regardless whether or not the Parthenon's architecture was built accordingly to the Golden Ratio, it is still an amazing structure, and may get some of its beauty from the regular rhythms introduced by the repetition of the same column (Livio, 2001). Renaissance artists often used diagonals and other interior lines of rectangles to divide rectangular space proportionally. For example the main diagonals of a rectangle allow for division of the rectangle into halves, both vertically and horizontally. Continuing, the diagonals of the halves allow division into quarters. Another tactic used by Renaissance artists to construct they work was called rabatment. Rabatment is where the shorter sides of the picture rectangle are rotated onto the longer. The rotation creates vertical division and overlapping squares. If rabatment is applied to a Golden Rectangle, the diagonals of the two overlapping squares cut the diagonals of the rectangle in golden proportion (Brinkworth and Scott, 2001). A construction of division by diagonals is provided on the left and a construction by rabatment is provided on the right (Figure 9.9).

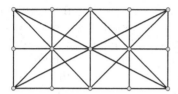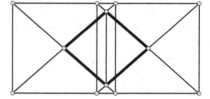

Figure 9.9

In the thirteenth century three artists' work contain close proportions to the Golden Rectangle. Italian painter and architect Giotto di Bondone (1267–1337) painted the "Ognissanti Madonna"[21] which is also known as "Madonna in Glory." Both the painting as a whole and the central figures in the painting can be inscribed by Golden Rectangles. Similarly, Sienese artist Duccio di Buoninsegna's (1255–1319) "Madonna Rucellai" and Florentine painter Cenni de Pepo's (1240–1302) "Santa Trinita Madonna" can be inscribed by Golden Rectangles. Both of these paintings are in the same room as the "Madonna in Glory." It is speculated that these three artists did not include the Golden Section into their paintings; rather they were driven by the unconscious aesthetic properties of the Golden Ratio. With respect to the time period, the three Madonnas were painted centuries before the publication of "The Divine Ratio" which brought the proportion into common knowledge (Livio, 2001).

Leonardo da Vinci inevitable comes into the discussion of the Divine Ratio and art. Five of his works have been speculated to host Golden Ratio properties: The unfinished canvas of "St. Jerome," the two version of "Madonna on the Rocks," the drawing of "a head of an old man," and the most famous of all, the "Mona Lisa"(Livio, 2001).

The two versions of "Madonna on the Rocks" have an interesting history. The first version, produced between 1483 and 1486, was done before da Vinci had any contact with Pacioli or his book "The Divine Ratio." The second version, which was completed around 1506, could have been influenced by Pacioli's book. Interestingly, both versions are very close to the Divine Ratio. In the first version, the dimensions are in proportion 1.64

and in the second version's dimensions are in proportion 1.58, both close estimates of ϕ (Livio, 2001).

Leonardo da Vinci's "head of an old man," is suggested to be a self-portrait which is overlaid with a square that is divided into rectangles. Some of these rectangles approximate Golden Rectangles but it is difficult to be absolutely sure. The rectangles are very roughly drawn and do not have square corners (Markowski, 1992). This suggests that depending on where one measures from, it is very possible to find some ratio that approximates the Golden Ratio. Leonardo da Vinci's "St. Jerome" has similar uncertainty. When overlaid with a Golden Rectangle, the left side of St. Jerome's body and his head are missed completely. The left side of the Golden Rectangle is tangent to a small fold of fabric and does not touch the body at all. Again, Leonardo was not introduced to Pacioli's book until thirteen years after the completion of "St. Jerome" (Markowski, 1992). His right arm also extends beyond the rectangle's side. The drawing of "a head of an old man,"[22] completed in 1490, is the closest demonstration that da Vinci used Golden Rectangles to determine dimensions in his paintings (Livio, 2001).

Human body proportions and facial features share similar mathematically proportioned relationships as other living organisms. The placement of facial features yields the classic proportions used by both the Romans and Greeks. Marcus Vitruvius Pollio described the height of a well proportioned man is equal to the length of his outstretched arms. The body and outstretched arms can be inscribed in a square, while the hands and feet are inscribed in a circle. With this system, the human body is divided into two parts at the naval. These parts are represented in the proportion of the Golden Rectangle. Classical statues from the fifth century such as Doryphoros the spear bearer and Zeus have the proportions suggested above (Elam, 2001).

The art described above deal with proportions of measurements. It should be noted that measurements, no matter how accurate, only provide reasonable estimates of the Golden Ratio (Fischler, 1981). The artist, painter or sculptor may or may not have been trying to conform to the proportion of the Golden Ratio. However close the approximations are, they could have been created with beauty in mind and with no intention to match the Golden Ratio.[23]

Visually pleasing art is not the only form of art where the Golden Ratio can be found. Music and mathematics have been entwined since antiquity and it is not surprising that one accompanies the other.[24] The Golden Ratio is related to many forms of music. Many listeners, including people who are only casually acquainted with the music of Mozart (1756–1791), can pick up on the manifested form and balance the composer used when writing his music (Putz, 1995).

Mozart worked with mathematical figures throughout his life. In his early composing years, he took up the problem of composing minuets 'mechanically', by putting two-measure melodic fragments together in a specific order. By the age of nineteen, Mozart had composed his first sonata for piano.[25] Almost all of his sonatas were composed of two movements: 1) the Exposition in which the musical theme in introduced and the Development and Recapitulation in which the theme is developed and revisited (Newman, 1963). A visual representation of Mozart's sonata-form movement can be seen below.

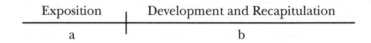

Sonata-Form Movement

The first movement of the first sonata, K. 279, is 100 measures in length. It is divided so that the Development and Recapitulation section has a length of 62. It should be noted that the lengths of the movements are natural numbers because they measure counts. When reviewing the first movement of the first sonata, it can be seen that 100 cannot be divided any closer (using natural numbers) to the Golden Ratio than 38 and 62. This is true for the second sonata which has total length of 74 and is divided in 28 and 46. A table of some of Mozart's movements is listed below.

Piece and movement	a	b	a + b
279, I	38	62	100
279, II	28	46	74
279, III	56	102	158
280, I	56	88	144
280, II	24	36	60
280, III	77	113	190
281, I	40	69	109
281, II	46	60	106
282, I	15	18	33
282, III	39	63	102

To evaluate the consistency of the ten proportions listed above, a scatter plot of b against a + b can be used (Figure 9.11). If a composer, Mozart in this case, is consistent with using the Golden Ratio in their works, the data should be linear and fall near the line y = φx. The graph on the left (Figure 9.11a) represents the degree of consistency by plotting the value of b with the values of a + b. The statistical analysis for the data shows an r^2 val-

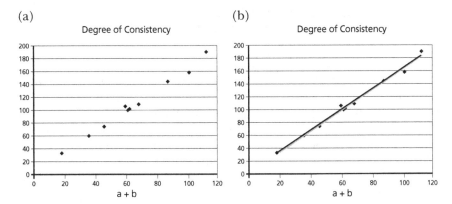

Figure 9.11

ue of .994 which confirms an extremely high degree of linearity. The graph on the right (Figure 9.11b) shows the linear regression of the data (represented by the gray line and the equation y = 1.59614205x + 2.733467326), and the line y = φx (black line) overlaid on the plot of the data. The statistical analysis of the data and the graphs below show that the data is linear and the points scarcely differ from the line y = φx. This is of impressive evidence that Mozart did partition sonata movements near the Golden Section (Putz, 1995).

If a movement is divided into the Golden Section, then both a/b and b/(a + b) should be near phi. Fischer (1981) provides a theorem and the following proof that b/(a+b) is always closer to φ than a/b is.

Theorem: $\{b/(a + b)\} - \phi \le (a/b) - \phi$ where $0 \le a \le b$.

Proof: Let x = a/b. Then show that,

$$\{1/(x)\} - \phi \le (x) - \phi$$

for all x [0,1]. Let f(x) = 1/(x + 1). By the Mean Value Theorem, for all x [0,1] there is a z (0,1) such that:

$$f(x) - f(\phi) = f'(z) \quad x - \phi .$$

Now $f'(x) = -1/(x + 1)^2$ satisfies

$$¼ < f'(x) < 1$$

For x (0,1). A simple calculation will show that φ is a fixed point of f, that is, that f(φ) = φ. So, for all x [0,1],

$$\{1/(x + 1)\} - \phi \ge (x) - \phi$$

with equality when x = φ. This theorem says that the ratio of consecutive terms of any Fibonacci-like sequence ($f_1 = a$, $f_2 = b$, $f_{n+2} = f_n + f_{n+1}$ with a and b not both zero) converges to φ.

MODERN IMPLICATIONS OF THE GOLDEN RATIO
AND BEAUTY

Beauty has been defined in many different ways since antiquity. A modern definition of beauty is "excelling in grace or form, charm or coloring, qualities which delight the eye and call forth that admiration of the human face in figure or other objects." Facial harmony can be activated through symmetry. Such symmetry exists when one side of the face is a mirror image of the other. The ideal face can be measured in symmetrical proportions. It should be noted that attractive faces are relatively symmetrical but not all symmetrical faces are considered beautiful (Adamson & Galli, 2003).

The Golden Ratio can also be found in human DNA structure[26] and has been found to be the only mathematical configuration that can duplicate itself ad infinitum without variance. It has been suggested that this represents a geometrically encoded instructional pattern in the brain that guides humans to recognize beauty.

The Golden Proportion can be found throughout a beautiful human face. The human head forms a Golden Rectangle with the eyes at the midpoint. The mouth and nose can each be placed at Golden Sections of the distance between the eyes and the bottom of the chin. With this information it is possible to construct a human face with dimensions exhibiting the Golden Ratio. This is exactly how some modern plastic surgeons are creating beauty. Dr. Stephen Marquardt created a Golden Decagon Mask, which is a two-dimensional visual perception of the face that has triangles with sides with ratios of 1:1.618. The Golden Decagon Mask is completed when forty-two secondary Golden Decagon matrices[27] are mathematically and geometrically positioned in the primary framework. The secondary matrices are geometrically locked on to the primary matrix by having at least two vertex radials, a vertex radial and an intersect of two vertex radials, or two intersects of vertex radials in common with the primary Golden Decagon matrix. These secondary Golden Decagon Matrices form the various features of the face (Marquardt, 2002). Figure 9.12 illustrates some examples of how the Golden Ratio is perceived throughout history and through different cultures.

Regardless of how the human face seems to fit into a unique geometric figure, beauty will always be defined in more ways than one. Plastic surgeons may construct beautiful faces today to fit into a Golden Decagon, but this may not always be the case. The future may lead to a new definition of

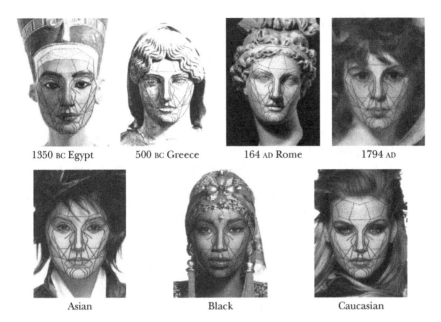

| 1350 BC Egypt | 500 BC Greece | 164 AD Rome | 1794 AD |

| Asian | Black | Caucasian |

Figure 9.12

beauty based on other information than the golden ratio. But it does make you wonder if "beauty is in the phi of the beholder."

CONCLUDING THOUGHTS

Phi could be the world's most astonishing number. It can be found in nature, throughout history, in art, music, and architecture. Many conflicting theories exist about the origins of phi (ϕ); however we cannot deny the principles that accompany it. Whether it is the mathematical relationships that seem to form around the number or the sheer aesthetics of the proportion, we must be aware that ϕ is all around us and rightly called the Divine Ratio.

ACKNOWLEDGEMENTS

The author wishes to thank Professor Bharath Sriraman for his valuable input on various drafts of the paper.

Reprint of Fett, B. (2006). An in-depth investigation of the divine ratio. *The Montana Mathematics Enthusiast,* *3*(2), 157–175. © The Montana Mathematics Enthusiast.

REFERENCES

Ackermann, E. (1895). The golden section. *The American Mathematical Monthly,* *2*(9/10), 260–264.

Adamson, P. & Galli, S. (2003). Modern concepts of beauty. *Plastic Surgical Nursing,* *24*(1), 32–36.

Amma, S. (1979). *Geometry in ancient and medieval India.* Delhi: Motilal Banarsidass.

Brinkworth, P. & Scott. P. (2001). The last supper at Milan. *Australian Mathematics Teacher,* *57*(3), 2–5.

Dunlap, R. (1997). *The golden ratio and Fibonacci numbers.* Singapore: World Scientific.

Elam, K. (2001). *Geometry of design: Studies in proportion and composition.* New York: Princeton Architectural Press.

Fischer, R. (1981). How to find the "golden number" without really trying. *Fibonacci Quarterly, 19,* 406–410.

Gupta, R. (1976) Sine of eighteen degrees in India up to the eighteenth century, *Indian Journal of History of Science, 11*(1), 1–10.

Herz-Fischler, R. (1987). *A mathematical history of the golden number.* Mineola, NY: Dover Publications, Inc.

Huntley, H. E. (1970). *The divine proportion: A study in mathematical beauty.* New York: Dover Publications.

Johnson, A. (1999). Fiber meets Fibonacci. *Mathematics Teaching in the Middle School, 4*(4), 256–262.

King, A. H. (1976). *Mozart in retrospect.* Westport, CT: Greenwood Press.

Livio, M. (2002). *The golden ratio: The story of Phi, the world's most astonishing number.* New York: Broadway Books.

Livio, M. (2003). Searching for the golden ratio. *Astronomy, 31*(4), 52–58.

Markowsky, G. (1992). Misconceptions about the golden ratio. *The College Mathematics Journal, 23*(1), 2–19.

Marquardt, S. (2002). Golden decagon and human facial beauty. *Journal of Clinical Orthodontics, 36*(6), 339–347.

Newman, W. (1963). *The sonata in the classical era.* Chapel Hill: University of North Carolina Press.

Niklas, K. & O'Rourke, T. (1982). Growth patterns of plants that maximize vertical growth and minimize internal stresses. *American Journal of Botany, 69*(9), 1367–1374.

Putz, J. (1995). The Golden Section and the Piano Sonatas of Mozart. *Mathematics Magazine, 68*(4), 275–282.

Wasler, H. (1996). *The golden section.* Washington DC: The Mathematical Association of America.

NOTES

1. Phidias was an Greek sculptor who lived between 490 and 430 BC. His sculptors included "Athena Parthenos" which is located in Athens and "Zeus" which is located in the temple of Olympia.
   Comment: astonishing is a strange word to use here . . . how about great?
2. Logarithmic spirals have a unique property. Each increment in the length of the shell is accompanied by a proportional increase in its radius. This implies that the shape remains unchanged over time and growth. As a logarithmic spiral grows wider, the distance between its coils increases and it moves away from its original starting point (pole). It turns by equal angles and increases the distance from the pole by equal ratios.
3. Golden Triangles are isosceles triangles that exhibit base angles of 72 degrees and an apex angle of 36 degrees. From the Pythagoreans and the construction of the pentagram (which has five equal-area golden triangles) it can be seen that the length of the longer side to that of the shorter side is in golden proportion.
4. A gnomon is a portion of a figure which has been added to another figure so that the whole is of the same shape as the smaller figure.
5. Mathematicians, Harold S., M. Coxeter, and I. Adler, showed that buds of roses which were placed in union with spirals generated by the Golden Angle were the most efficient. For example, if the angle used was $360/n$ where n is an integer, the leaves would be aligned radially along n lines, thus leaving large spaces. Using the Golden Angle, which is an irrational multiple of 360 degrees, ensures that the do not line up in a specific radial direction and this leaves no space unfilled.
6. Fibonacci is a shortened form of Filius Bonaccio (son of Bonaccio). Fibonaci was taught the Arabic system of numbers in the 13th century. He later published the book *Liber Abaci* (Book of Abacus). This book introduced the Arabic numbering system to Europe and gave Fibonacci everlasting fame as a mathematician (Dunlap, 1997).
7. Fibonacci Numbers are represented by the recursive relation $A_{n=2} = A_{n+1} + A_n$.
8. The Platonic Solids Plato used consisted of five shapes. The first three; tetrahedron, octahedron and the icosahedron, were based on equilateral triangles. The remaining two; cube and dodecahedron were made from the square and regular pentagram.
9. Plato's theory was much more than a symbolic association. He noted that the faces of the tetrahedron, cube, octahedron, and dodecahedron could be constructed out of two types of right angled triangles, the isosceles 45–90–45 and the 30–60–90 triangle. Plato explained that his chemical reactions could be described using these properties. For example, when water is heated by fire, it produces two particles of vapor (air) and one particle of fire, {water} → 2{air} + {fire}. In Platonic chemistry, balancing the number of faces involved (in the Platonic solids that represent these elements) we get $20 = 2 * 8 + 4$. The central idea is that particles in the universe and their interactions can be described by a mathematical' theory that possesses certain symmetries.

10. In general a regular n-gon has n edges and interior angles given by the equation $\alpha = [1 - (2/n)] * 180$.

11. We have to place certain restrictions on the values of m. These reasons are if $m = 2$ then an edge is formed, not a vertex. And if $m\alpha = 360$ degrees, then the vertex is merely a point on a plane and if $m\alpha > 360$ degrees then the faces overlap.

12. The table above lists the characteristics of the five Platonic Solids. The quantities n and m are the number of edges per face and the number of faces per vertex. The quantities e, f, and v are the total number of edges, faces, and vertices for the respective solid.

13. Pythagoras emigrated to Croton in southern Italy sometime between 530 and 510. He studied Egyptian, and Babylonian mathematics, but both of these prove too applied for him. There are many different accounts of the Mathematician's life and death, but what is known for sure is that he was responsible for mathematics, and philosophy of life and religion.

14. The pentagram is closely related to the regular pentagon. If one is to connect all the vertices of the pentagon by diagonals, a pentagram is constructed. The diagonals of this pentagon form a smaller pentagram. This process can be continued to infinity, and every segment is smaller that its predecessor by a factor that is precisely equal to the Golden Ratio.

15. Side and diagonal numbers of squares start off with the number one as the first number in the sequence. For pentagonal side and diagonal numbers, starting with one will lead to the degenerate case. Thus we have to start with the two as the first number in the sequence.

16. The 8:5 triangle was an isosceles triangle in which the base was eight units and the height was five units.

17. The Golden Rhombus is a two dimensional figure that has perpendicular diagonals which have a ratio of 1:ϕ.

18. Athena is the Greek goddess of wisdom, war, the arts, industry, justice and skill. Her father was Zeus and her mother was Metis, Zeus' first wife.

19. The Parthenon, know as "the Virgin's place in Greek," in Athens was built in the fifth century BC and is one of the world's most famous structures. The Parthenon is a sacred temple to the cult of Athena Parthenos.

20. On September 26, 1687, Venetian artillery directly hit the Parthenon. General Konigsmary said "How it dismayed His Excellency to destroy the beautiful temple which had existed for over three thousand years."

21. Bondone's painting "Madonna in Glory" is currently in the Uffizi Gallery in Florence. This painting features an enthroned Virgin with a child on her lap. Both Madonna and Child are surrounded by angles.

22. The drawing of "a head of an old man" is currently in the Galleria dell' Accademia in Venice.

23. Fischler (1981) gives a detailed description, complete with proofs of how certain data can be transformed to exhibit Golden Ratio characteristics.

24. When Mozart was learning arithmetic, he gave himself entirely to it. His sister recalls that he once covered the walls of the staircase and of all the rooms in their house with figures, then moved to the neighbors house as well (King, 1976).

25. Mozart wrote 19 all together.

26. DNA molecules are based on the Golden Ratio. A single DNA molecule measures 34 angstroms long by 21 angstroms wide for a full cycle of its double helix spiral. Both 34 and 21 are Fibonacci numbers which converge to the Golden Ratio. The double-stranded helix DNA molecule has two grooves in its spiral. The major groove measures 21 angstroms and the minor groove measure 13 angstroms, again, both are Fibonacci numbers. Another unique way that DNA is related to the Golden Number can be seen in a cross-sectional view of a DNA strand, which turns out to be a decagon. The golden properties of the decagon are discussed above.

27. The secondary Golden Decagon matrices are constructed exactly the same way as the primary Golden Decagon only smaller.

CHAPTER 10

CYCLIDE MANIPULATION

Akihiro Matsuura
Tokyo Denki University (Japan)

ABSTRACT

A Dupin cyclide is a non-spherical algebraic surface discovered by French mathematician Pierre-Charles Dupin (1822). We use it as what we call a visual instrument, that is, an instrument which enables image-provoking movement in space through manipulation. Then, mathematical curves such as cycloids and trochoids are approximately drawn as loci of points on the moving cyclide. We also develop some illusional movement which are equal to the basic mime illusions. We constructed some Dupin cyclides and created a performance in which image-provoking nature of a cyclide is highlighted. Finally, we discuss the possibility of introducing the ideas of manipulating geometric objects in mathematics learning.

INTRODUCTION

Since old times mathematical ideas have been used for creating art works. Da Vinci, Dürer, and Escher, among others, have used mathematical ideas effectively in their work. A connection between art and mathematics is seen not only in static types of arts such as painting, sculpture, and architecture but also in dynamic ones such as kinetic and performing arts.

Interdisciplinarity, Creativity, and Learning, pages 133–143

133

Juggling is one of the performing genres to throw several objects in the air and catch skillfully using hands. Recently, it has been also explored mathematically and artistically. Mathematically, Shannon, who is the founder of information theory, proved a theorem which determines the timing of catching and throwing of objects (Shannon, 1980). In 1980's a mathematical notation for expressing juggling patterns was discovered independently by three groups of people (Tiemann & Magnusson, 1989; Polster, 2002). Furthermore, some fundamental theorems concerning juggling patterns were proved (Buhler, Eisenbud, Graham, & Wright, 1994; Polster, 2002). From an artistic standpoint, performing artist Michael Moschen made pioneering work by creating novel performances through exploration of the nature of moving objects. For example, at the performance called *Triangle*, he bounces a set of balls in a huge triangle and makes a variety of patterns and rhythms. Greg Kennedy rolls multiple balls in a clear hemisphere and draws beautiful patterns. Recently, the author presented a new kind of juggling called *spherical juggling* (Matsuura, 2004), in which balls are inserted in a large clear globe and rolled on the inner surface. We note that in all of these performances balls are not necessarily thrown in the usual sense but are manipulated in novel ways using simple geometric objects.

In this paper we present yet another method of manipulating mathematical objects. Namely, we use a geometric surface called a *Dupin cyclide* and explore its novel usage. A Dupin cyclide is a non-spherical algebraic surface discovered by French mathematician Pierre-Charles Dupin (Dupin, 1822). It has the following properties desirable for manipulation. First, a cyclide has a remarkable property that all the lines of curvature are circular. This affects to its simplicity, beauty, and easiness in the handling. Furthermore, we can observe beautiful curves such as trochoids, e.g., *cycloids*, *curtate cycloids*, *prolate cycloids*, *epitrochoids*, and *hypotrochoids*, approximately as loci of points on the moving cyclide. We also find some illusional movement of a cyclide which makes us feel that the cyclide moves autonomously. Lastly, the shape of cyclides is so universal that we can find similar shapes in many fields, e.g., as a crescent, a croissant, a pair of horns, a bag, a lock, and etc. All of these properties imply that a cyclide can be an attractive visual instrument which is image-provoking and is used for artistic expression.

We constructed Dupin cyclides and explored the method to manipulate them. Then, we created a performance in which the properties of cyclides were fullly utilized. We report on our experience of performances and finally, we make a brief discussion on introducing the idea of manipulating geometric objects in mathematics learning.

SUMMARY OF GEOMETRIC CONCEPTS

Annuli

The area lying between two concentric circles is called an *annulus* and is shown in Figure 10.1. More generally, the area lying between two circles in Figure 10.2 are called a *shifted annulus*. We note that in the actual construction of an object, it has some thickness and volume. However, we abuse the name annulus in this case. The motivation to consider a shifted annulus is that such a form is seen in many fields and quite image-provoking. Furthermore, asymmetry of a shifted annulus contributes to generating unexpected locus. Details will be explained later.

A Dupin Cyclide

There are several methods to define a Dupin cyclide. It is in one way defined to be the image of a circular cylinder or a torus under an inversion. It is also defined to be the envelope surface of spheres tangent to three fixed spheres (Dupin, 1822). A ring cyclide, which is a special case of a Dupin cyclide, is shown in Figure 10.3. In this paper, we focus on this type of cyclide.

Figure 10.1 An annulus.

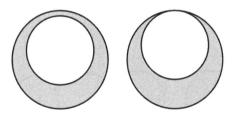

Figure 10.2 Shifted annuli.

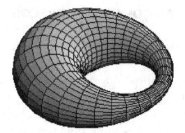

Figure 10.3 A Dupin cyclide (a ring cyclide).

A Dupin cyclide has a remarkable property that all the lines of curvature are circular. This contributes to its simplicity, beauty, and easiness in the handling. At a performance, we want an object to be simple enough to be handled well and expressive enough to produce a variety of rich visual images. A Dupin cyclide satisfies this criterion.

The Family of Cycloids and Trochoids

When a curve C' rolls on a fixed curve C without sliding, the locus of point P on C' is called the *roulette* for P. C is called the *base curve* and C' is called the *rolling curve*. In the case that C is a straight line and C' is a circle, the roulette is called a *cycloid*. In Figure 10.4, when the circle rolls on the straight line, the point P_1 on the circle draws a cycloid. It is possible to think of a more general setting that a traced point is off the rolling curve. In this case, the locus is called a *trochoid*. Cycloids and trochoids are written with polar coordinates as $x = a\theta - b\sin\theta$ and $y = a - b\cos\theta$ (Weisstein). Here, a is the radius of the base circle and b is the distance between the traced point and the center of the circle. If $b < a$, the trochoid for point P_2 in Figure 10.4 is also called a *curtate cycloid*. If $a < b$, the trochoid for point P_3 in Figure 10.4 is also called a *prolate cycloid*. It might be interesting to note that curtate cycloids have been used by some violin makers for cross arching of musical instruments and they resemble the arching of some great Cremonese instruments such as those by Stradivari (Playfair, 1999).

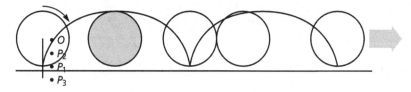

Figure 10.4 A rolling circle and a cycloid for P_1.

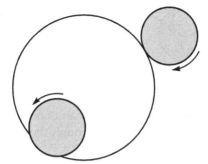

Figure 10.5 Rolling circles outside and inside the base circle.

If some of the points on and off the circle are visually emphasized in some way, the family of cycloids can be observed simultaneously when the circle itself goes straight. Such a phenomenon will be actualized with a cyclide in the next section.

Next, we review the case that the rolling curve and the base curve are both circles. There are two possible cases. The rolling circle is located (i) outside the base circle, or (ii) inside the base circle. The rolling circles are shown in Figure 10.5. In case (i), the roulette made by a point on the rolling circle is called an *epitrochoid*. In case (ii), the roulette made by a point on the rolling circle is called a *hypotrochoid*. In more detail, suppose that the radius of the rolling circle is b and the radius of the base circle is a. In case (i), if $a = b$, the roulette of a point on the rolling circle is a *cardioid*. If $a = 2b$, the roulette of the point is a *nephroid*. In case (ii), the roulettes of a point on the rolling circle also differ according to the values of a and b. They are called *spirographs*. All of these curves are attractive in an aesthetic point of view. For the detail of these curves, see for example the website MathWorld (Weisstein).

ANNULUS AND CYCLIDE MANIPULATION

Rolling on a Straight Line

We now consider the movement of a shifted annulus and a cyclide. Since these objects are identified through the orthogonal projection, we may focus on a shifted annulus for basic two dimensional movement. First, we roll the annulus on a straight line without sliding. In this case, since the outer boundary of the object is a circle, cycloids naturally appear as the roulettes of points on the outer circle as shown in Figure 10.6. On the other hand, a point in the outer circle, especially, the center of the circle, draws a curtate cycloid. This means that the upper and lower envelopes of the inner

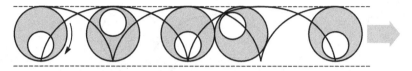

Figure 10.6 Straight lines as envelopes and the drawn cycloids.

circle are curtate cycloids, too. So, we can observe straight lines, cycloids, and curtate cycloids simultaneously through the manipulation of a shifted annulus or a cyclide.

At a performance, a base curve might not be set explicitly. In such a case, we imagine the curve in space and so, the mathematical loci such as cycloids are actualized approximately. In the manipulation, the object is handled in various ways. A natural and dynamic way is to insert one hand in the inner circle and roll the object while moving it straight as in Figure 10.6. When both hands are used, the movement of the object would be more stable.

Next, prolate cycloids are observed as follows. We determine the inner circle to roll on a straight line without sliding as shown in Figure 10.7. Then, each point on the outer circle draws a prolate cycloid. The similar movement is possible with a normal annulus, however, the amplitudes of prolate cycloids are much bigger and perceived more clearly with a shifted one.

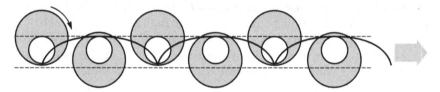

Figure 10.7 Straight lines as envelopes and the drawn cycloid.

Rolling on a Circle

Next, we roll a shifted annulus along a circle. There are mainly two types of manipulation: (i) One is to roll the inner or outer circle on the outer boundary of the base circle (Figure 10.8); and (ii) the other is to roll the inner or outer circle on the inner boundary of the base circle (Figure 10.9). In case (i), a family of epitrochoids is observed simultaneously. On the other hand, in case (ii), a family of hypotrochoids is observed.

In the actual manipulation, we draw curves which are not necessarily mathematical. We may even slide the object to fully express its nature. Still, mathematical concepts such as straight lines, circles, cycloids, and trochoids are used as the basis for creating a variety of images.

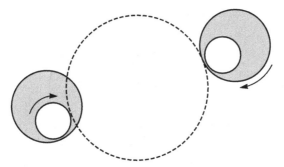

Figure 10.8 Rolling the inner or outer circle on the base circle to draw epitrochoids.

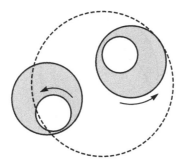

Figure 10.9 Rolling the inner or outer circles on the base circle to draw hypotrochoids.

Illusional Movement

We now illustrate some illusional techniques using an annulus and a cyclide. First, we insert our hand to the inner circle of the object. The black point in Figure 10.10 is where the hand is located. Suppose that there is little friction between the hand and the surface of the object. Then, the object tends to move down along the tangent line. Now, what if we pull the hand *upper-backward* as shown in Figure 10.10? Interestingly, though the force is given backward, the object proceeds *forward*. This phenomenon is explained with the composition of the two vectors as shown in Figure 10.10. If the object is a cyclide and the friction is very little, we may grasp the tubular part of the cyclide around the black point instead of inserting a hand to actualize the same movement. This somewhat illusional movement makes us feel that the object has its own life.

Another type of illusional movement is possible when the object is manipulated as follows. The black point in Figure 10.11 is again where a hand

pull the object

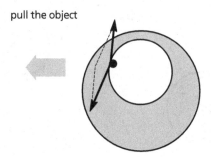

Figure 10.10   The object proceeds forward as it is pulled upper-backward.

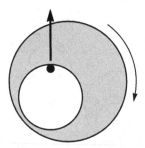

Figure 10.11   The object rotates clockwise as the hand is raised upward.

is located. Then, if we raise the hand upward, the object begins to rotate clockwise as shown in Figure 10.11. It is physically natural, but it has an illusional visual effect because the manual straight movement is automatically changed to a rotation. Similarly to the movement in Figure 10.10, this phenomenon also makes us give an autonomous character to the object. The similar technique is developed by moving the hand horizontally, which results in rotating the object. Combination of these techniques makes a variety of illusional movement.

CONSTRUCTION AND PERFORMANCE

Construction

To verify the expressiveness of a shifted annulus and a Dupin cyclide as a visual instrument, we constructed them and explored the techniques possible with the actual objects. The sizes of the objects we constructed are as follows. The diameters of the outer and inner circles of the objects we constructed are 13–18 inches and 6–9 inches, respectively. The diameters of the circles of the finest and fattest parts of the cyclides are 1 and 6–8 inches,

Figure 10.12   A constructed Dupin cyclide.

respectively. The annuli are made of wood and the cyclides are made of FRP (Fiber Reinforced Plastics). One of the constructed cyclides is shown in Figure 10.12.

Concepts

The main aim of this research is to find a new connection between mathematics and performing arts. Using the cyclide we constructed, we first developed techniques for manipulating it. Some of the image shots of the cyclide manipulation are shown in Figure 10.13. Then, we created a performance based on the unique movement of the object and the character which is imagined from the appearance and the movement. This time we gave it a character of a capricious moon (a crescent). The rough sketch of the performance is written as follows.

> The moon cyclide is shadowy and gleaming in light. It is hanged solely in space. A performer wanders and happens to find it. The appearance is so attractive that he approaches and holds it. Then, the cyclide suddenly begins to move away from him as if it has its own life. The performer follows. The cyclide floats back and forth, up and down, and rotates freely to draw beautiful curves in space. After playing for a while, the cyclide returns to its place.

Performances

In the fall of 2004, we attended the International Performance Festival in Shizuoka, Japan as a performer and presented the cyclide performance

Figure 10.13 Images of the cyclide manipulation.

several times. The music we used is the classical music "Clair de lune" by Debussy. Every time there were about 150–200 people as audience. Though the stage was outside, the performances were done in a concentrated atmosphere. The appearance and the movement of a cyclide and its connection with the performer seemed to attract the audience. However, some said that it was difficult to recognize that the object was meant to be a moon. Not a ring cyclide but a horn cyclide might be better. The refinement of the techniques and the act is also necessary to fully express the fantastic atmosphere of the performance.

CONCLUDING REMARKS

In this paper, we presented a method of using a Dupin cyclide and a shifted annulus to create a new kind of performance. The benefits of this paper is summarized as follows.

- A new connection between mathematics and performing arts is presented. Especially, a manipulative use of a Dupin cyclide and exploration of curves drawn by a cyclide would be the key advance.
- Some illusional movement of a cyclide was. It might be interesting to note that such a movement was never imagined before actually constructing and manipulating the object. Namely, it was an example that substantiation and tangibilization were crucial to develop an idea.
- The idea of manipulating geometric objects can be introduced in mathematics learning because real objects and their dynamic ma-

nipulation help learners to visualize and tangibilize mathematical concepts. Furthermore, such a dynamic manipulation also brings liveliness to a class.

Still, there are many problems to deal with. As for the cyclide, refinement of the product, techniques and the performance should be done. Exploration of other geometric surfaces would also be the future work. From the educational standpoint, we need to think of the dynamic and tangible ways for learning the mathematical concepts which appear in the usual courses. To work in close cooperation with dynamic geometry software as well as the actual objects is also the next step.

ACKNOWLEDGEMENTS

This work is supported in part by the Grant of Hayao Nakayama Foundation for Science, Technology and Culture.

Reprint of Matsuura, A. (2005). Cyclide manipulaton. In Beckmann, A., Michelsen, C., & Sriraman, B (Eds.). *Proceedings of the 1st International Symposium of Mathematics and its Connections to the Arts and Sciences.* The University of Education, Schwäbisch Gmünd, Germany, pp. 99–107.

REFERENCES

Buhler, J., Eisenbud, J., Graham, R. L., & Wright, C (1994). Juggling drops and descents. *American Mathematical Monthly, 101*(6), 507–519.

Dupin, P.-C (1822). Applications de géométrie et de méchanique: à la marine, aux ponts et chaussées, etc. *Bachelier*, Paris.

Matsuura, A (2004). Spherical juggling. *Proc. of Interdisciplinary Conference of the International Society of the Arts, Mathematics, and Architecture (ISAMA/CTI 2004)*, 89–94.

Playfair, Q (1999). Cremona's Forgotten Curve. *The Strad, 110,* 1194–1197.

Polster, B (2002). *The Mathematics of Juggling*, Springer Verlag.

Shannon, C. E (1980). Scientific Aspects of Juggling. *Claude Elwood Shannon: Collected Papers*, edited by Sloane, N. J. A. and Wyner, A. D., IEEE Press, 850–864 (1993).

Tiemann, B. & Magnusson, B (1989). The Physics of Juggling, *The Physics Teacher*, 27, 584–589.

Weisstein, E. W. Trochoid, *MathWorld—A Wolfram Web Resource,* from http://mathworld.wolfram.com/Trochoid.html

SECTION IV

INTERDISCIPLINARITY AND MODELING

CHAPTER 11

MODELING INTERDISCIPLINARY ACTIVITIES INVOLVING MATHEMATICS AND PHILOSOPHY

Steffen M. Iversen
University of Southern Denmark

ABSTRACT

In this paper a didactical model is presented. The goal of the model is to work as a didactical tool, or conceptual frame, for developing, carrying through and evaluating interdisciplinary activities involving the subject of mathematics and philosophy in the high schools. Through the terms of Horizontal Intertwining, Vertical Structuring and Horizontal Propagation the model consists of three phases, each considering different aspects of the nature of interdisciplinary activities. The theoretical modelling is inspired by work which focuses on the students abilities to concept formation in expanded domains (Michelsen, 2001, 2005a, 2005b). Furthermore the theoretical description rest on a series of qualitative interviews with teachers from the Danish high school (grades 9–11) conducted recently. The special case of concrete interdisciplinary activities between mathematics and philosophy is also considered.

Interdisciplinarity, Creativity, and Learning, pages 147–164
Copyright © 2009 by Information Age Publishing
147

INTRODUCTION

There is worldwide consensus that the society we live in today gets increasingly more and more complex. Earlier the problem was often to gather information, whereas the knowledge society of today is characterized by the fact that much information is easy accessible. The problem nowadays is therefore to survey and filter the great amount of accessible information rather than to gain access to it. Thus, the schools have to aim at producing students who are prepared to deal with such a great complexity of knowledge, that is, scientifically literate students (Gräber et al., 2001). In the educational system knowledge is still in a very large scale separated into distinct blocks by different subjects. This separation of knowledge has shown itself to be very efficient in producing and teaching new knowledge, but does not necessarily provide the students with the skills necessary to navigate through the constantly increasing amount of accessible information. Interdisciplinary activities between different subjects can help to develop a broader context of meaning or understanding for the student, and in this way contribute to the ongoing scholarly development and provide the student with the tools necessary to deal with complex problem solving waiting in the future.

In spite of the fact that many different subjects and areas often contain more and more mathematics-rich elements, mathematics, as a subject, mathematics remains quite isolated. The objective importance of mathematics from a social point of view exists side by side with its subjective irrelevance experienced by many people. Niss et al. (Niss, Jensen, Wedege, 1998) have characterized this as the *relevance-paradox* of mathematics. One reason for this could be found in the fact that mathematical knowledge is hard to transfer to new domains of knowledge by the student. Although the subject of mathematics in its very nature often is described as a tool, and therefore should be able to establish obvious connections to other contexts, such transfers of mathematical knowledge between different domains seldom occur (Hatano 1996, Michelsen 2001). Most of the time subjects have their own specific use of language and system of terminology and this can prevent the desirable transfer of mathematical knowledge to other contexts and domains.

The purpose of using interdisciplinary elements in the teaching of mathematics is, as concluded from the description above, 1) an attempt to broaden the students' curricular perspective and general view by removing the discrete lens that characterizes most schools' separation of knowledge into curriculums and present to the students a more real picture of the role and importance of mathematics in extra-mathematical contexts[1] 2) an attempt to help the students' abilities to transfer mathematical knowledge between different curricular domains.

To be able to work with interdisciplinary aspects in the teaching of mathematics one has to consider what connections the subject mathematics has to other subjects and areas of knowledge. In *the first International Symposium of mathematics and its Connections to the Arts and Sciences*[2] (Beckmann, Michelsen & Sriraman, 2005) such connections were discussed, and a sketch of a didactical model for interdisciplinary activities between mathematics and philosophy presented (Iversen, 2005).

Afterwards, the modeling of such activities involving mathematics has continued and the purpose of this paper is to present a didactical model, a conceptual frame for the planning, completion and evaluation of successfully interdisciplinary activities involving mathematics. The model will function as a tool to help develop activities that can facilitate a reasonable transfer of mathematical knowledge to other subjects and domains. The model is inspired by the work of Michelsen (2001, 2005a, 2005b) and is further developed through the special case of mathematics and philosophy and a section is therefore devoted to this specific topic. The section will also work as a demonstration of how the model should be understood and applied.

THEORETICAL FRAMEWORK

Working with interdisciplinary activities implies a belief that there exist elements that is general and somewhat identical between the knowledge presented in different subjects. We assume that such an intersection of knowledge contains more elements the more related[3] the subjects are to one another, and at least *not-empty* (Dahland, 1998). There are different ways of trying to describe such assumed curricular intersections. In the development of the didactical model presented here a notion of *competencies* is used to identify and characterize the possible intersection of knowledge between mathematics and other subjects.

In the educational system of Denmark a huge step forward is taken with the completion of the *KOM-report* for mathematics (Niss et al., 2002). In this Niss lists eight mathematical competencies, valid for all steps of education, which is a meant to work as an overall frame for description of the education in mathematics in Denmark. The concept of a mathematical competence is here understood as some sort of *mathematical expertise,* and is more formally defined as *an insightful readiness to act appropriate in situations which contains a certain kind of mathematical challenges.*[4] The report has been a starting signal to similar competence descriptions of other subjects in the Danish educational system.

A description of mathematics by the means of competencies focuses more on the *purpose* of learning mathematics than to the specific curriculum. This description expresses a broader minded view on the teaching of mathematics than a normal curricular-dependent view. But Niss describes

(Niss et al., 2002, p. 66) the eight mathematical competencies as strictly belonging to the sphere of mathematics thereby partly closing down the newly constructed bridge to other subject domains. Michelsen et al. (2005a) instead argues that some of the competences put forward by Niss et al (2002) are actually *interdisciplinary competences,* and mentions the modeling and representational competence as examples.

In this paper we will try to make use of the interdisciplinary potential inherent in a competence approach to mathematics on a theoretical didactical level suggested by Michelsen et al (2005a). A less bounded description of mathematical competences can then be substratum that enables an entanglement of mathematics with other subjects both on an educational theoretical level and on a practical level in the classrooms. It is here suggested that the notion of a mathematical competence should contain both a *narrow* and a *broad* dimension, by means of which such characterization of mathematical expertise in the student can both work as a description internally in mathematics and as a link to the rest of the world. As an example Niss (2002) mentions *the ability to reason mathematically* i.e., to be able to follow and judge mathematical argumentation, as one of the eight described competences. But the ability to be able to follow and judge a reasoning is far from restricted to the sphere of mathematics. It is the kind of expertise that is important to master in all the school's different subjects, and it could therefore be argued that some sort of *reasoning competence* is just as essential in physics or philosophy as it is in mathematics. Obviously arguments and reasoning often appear in different use of language and forms in different subjects, and therefore a reasoning competence is here suggested to be characterized by *the ability to follow and judge a reasoning in different curricular domains,* AND *being able to distinguish and characterize different types of arguments* thereby having the ability to go deeply into a certain subject and follow and judge a reasoning characteristic for this one subject.

Within mathematics valid arguments often have character of a proof, while arguments in other subjects, as e.g., philosophy or history, often are marked by less cogency and more contingent elements. In this context mastering the reasoning competence will be understood as the ability to distinguish different kinds of arguments but at the same time know why the different arguments work in different contexts, and to be able to dive into a specific argument, as e.g., a mathematical proof, and follow its string of reasoning.

This broad minded approach to the notion of competences should be understood as an attempt to, over time, change the educational practice which makes it possible that

> Although critical thinking, problem solving and communication are real world skills that cut across the aforementioned disciplines students are led to believe that these skills are context dependent. (Sriraman, 2004, p. 14)

INTERVIEWING HIGH SCHOOL TEACHERS

During May 2005 a series of qualitative interviews were conducted. Six high school teachers were interviewed individually. The main purpose was to find out: *Which didactical (and practical) possibilities and obstacles exist for interdisciplinary activities between mathematics and other subjects (especially philosophy) in the Danish high school (grade 9–11)?*[5] The interviewees were teachers from different high schools in Denmark and varied both in age and seniority. They were chosen so that each one taught either mathematics or philosophy (or both) on a daily basis and moreover most of them had been engaged in relevant interdisciplinary activities. The hope was to be able to incorporate some of this real-life information into the development of the didactical model. In the following I will reproduce some of the, for this paper, relevant conclusions one can draw from the conducted interviews.[6]

Some of the interviewed teachers have conducted interdisciplinary activities between mathematics (or physics) and philosophy earlier on in their daily teaching. It has not been possible to find any writings about conducted activities between mathematics and philosophy in the Danish high school, but some of the interviewed teachers have been involved in documented activities involving physics and philosophy. Generally the experiences from these courses were positive

> It's easy for me to register that the students have been going through these activities (involving physics and philosophy) and other teachers can easily do so to.... They [the students] own more academically concepts than students usually have. They are really good at thinking different subjects together, and they also get very good at working together in little groups...I think they simply have a greater cultural and historical horizon. (Teacher 1)

The purpose of these activities involving physics and philosophy was primarily to strengthen the subject of physics. To embody the abstractness of physics as one of the interviewed teachers told me. This goal was in some sense achieved according to the teacher quoted above and the reports of evaluation carried out by the involved students and teachers afterwards. Besides the registered positive cognitive effects the students realized that physics can not be reduced to a mere collection of dead facts. Physics is a human activity that evolves and therefore argumentation actually do count. This shift in the students' perception of the subject physics from being a dusty collection of facts, to being relevant, is an experience that another of the involved teachers believe can be re-produced in the case of mathematics.

> We can re-create the part about discovering in the case of mathematics...For example the students often only see the end-product when they see a proof for some mathematical relation. For them it's often a strange thing; How have

"they" found out you are supposed to do like that? They ask themselves. The process from a proof starts to crystallize and right to the final version of the proof which needs polishing before it appears in a textbook, nice and rounded. That whole process one should try in teaching mathematics, I believe it would be very beneficial for the students. (Teacher 2)

Besides using philosophy as a tool to illustrate the world and methods of physics the teachers involved report how at the same time the activities created the perfect interdisciplinary context for developing central concepts from the philosophy of science. Ideas such as: *induction, empirically investigations* and *verification* were easy for the students to acquire and work with in this expanded domain. In this way the activities held the possibility that both involved subjects could engage in the work of developing the students' scientific literacy, but at the same time use the cross-curricular context to discover and develop relevant aspects specific to the different curriculums.

Others of the interviewed teachers had themselves planned and conducted interdisciplinary activities involving mathematics and philosophy. In both cases the activities had been carried out in relation to the daily teaching of mathematics, and both set of activities centered about *argumentation and proof in mathematics.* The purpose of the different activities varied slightly but fundamentally they both tried to illustrate characteristics of mathematical argumentation and how this often is worked out.

When we speak about method, we did something about; When do you examine something and when do you actually construct a proof? And also, what is needed to construct a proof and what is the nature of a mathematical proof? These issues are very philosophical I think, and the activities were a great success for the students. (Teacher 3)

We worked with paradoxes and reasoning and things like that... The overall theme was argumentation. It was a very good course, and the students were very fond of it. (Teacher 4)

The work with these topics in mathematics was carried through based on a wish to equip the students with some general tools, or concepts, which could function as some sort of cognitive scheme for their ongoing daily struggle for learning mathematics.

A part of the teaching is about giving them [the students] a set of concepts which they can use to relate to what the are doing concretely. When they engage in a specific task in mathematics, they now have some concepts, some work habits, some patterns, some ways of thinking which they can use to throw light on what they are actually doing. (Teacher 3)

Mathematics propagates through a large and branching taxonomy of concepts and ideas. Several of the interviewed teachers pointed out that, cross-curricular activities between mathematics and a subject as philosophy should deal with concepts placed fairly high in the mathematical taxonomy used in the high school. To illustrate this point we can consider the relative position of two mathematical concepts in the taxonomy. Look for example at say the concept of *function* and a specific function as $f(x) = \sin(x)$. Both entities can be considered as a concept that a student in the high school should become acquainted with at some point. The concept of function however will be placed highest of the two in a taxonomy of mathematical concepts, and we will therefore regard this as a *meta-concept* in comparison with $f(x) = \sin(x)$. This way there also exists meta-concepts in comparison with the concept of function. The concept of *functional* is an example of a such, and the use of the name meta-concept will therefore always be relative.

For high school students the concept of *proof* will be regarded as a meta-concept most of the time and a direct investigation of this in the classroom by the students will often involve several problems. According to Dreyfuss (1999) most of the students on this educational level has a very restricted knowledge about what constitutes a mathematical proof. Also Hazzan and Zazkis (2005) point to the importance of trying to help the students acquire relevant mathematical meta-concepts as e.g., the proof.

According to Niss (1999) a major finding of research in mathematics education is students' alienation from proof and proving. Students' conceptions of the mathematical proof and those held by the mathematical community is separated by a huge gap. Niss concludes that

> Typically, at any level of mathematics education in which proof or proving are on the agenda, students experience great problems in understanding what a proof is (and is not) supposed to be, and what its purposes and functions are, as they have substantial problems in proving statements themselves, except in highly standardized situations. (Niss , 1999, p. 18)

Instead the students' consider proofs and proving as strange rituals performed by professional mathematicians that are not really meant to be understood by ordinary human beings. The activities referred to above by the interviewed teachers are exactly concerned with these problems and shows how other subjects such as philosophy can be used in the struggles.

The interviewed teachers generally believed that interdisciplinary activities involving mathematics were very relevant for the students. Focusing on the special case of mathematics and philosophy some of teachers suggested that relevant activities could take as a starting point the purpose of illuminating the structure of mathematics, its fields of study and its characteristic form of argumentation. It comes as no surprise that the examples mentioned here are of a very general character. Engaging in interdisciplinary activities should hold the possibility of gaining something for all the involved subjects, and this would indeed be a very difficult premise to fulfill for *both* mathematics and philosophy if the activities centered about the quadratic equation and Socrates' famous Defence. Both are examples of a far to narrow approach to interdisciplinary activities determined too much by curricular considerations.

In spite of a general optimism shared by the interviewed teachers towards integrating the teaching of mathematics with other subjects, several of them also point to a number of difficulties with the subject of mathematics that must be overcome if the interdisciplinary activities should be rewarding.

The subject of mathematics is regarded as a subject that holds great technical difficulties for the students. According to the interviewed teachers exciting problems and topics in mathematics often demands a severe amount of preparation from the students before they can engage with the activities thereby losing the immediate interest that is so important for the learning process (Mitchell, 1993). Other subjects, e.g., philosophy, is for most students easier to engage in and this often leads to a shift in the students attention away from the mathematical content of the chosen topic. For that reason the development of successful interdisciplinary activities involving mathematics needs the development of a working culture among teachers and students where it is respected that a subject as mathematics can be hard accessible and show this problem extra attention in the classroom.

Most of the interviewed teachers highlighted the fact, that in many cases interdisciplinary activities end up bringing in the mathematics teacher to simply help the students read of some values on a prefabricated curve or similar. Here the actual mathematical content is far from challenging or relevant for the students (or the teacher). To avoid this situation one of the interviewed teachers point out that

> There's an interaction between the other subject [than mathematics], the way it asks its questions and the areas of mathematics you can point out and work with. Sometimes mathematics and the other subject actually pose the same kinds of questions but they each give different kinds of answers. . . . The problems that the activities are meant to center on must have double-relevance, and that means that they should have relevance both in the reality to which they belong and also in mathematics. As a thought I think that is very correct because often they [the other teachers] say; Yes, this topic is really interesting could the mathematics teacher please come in here and help reading of the curve! I answer: No, no that's not really interdisciplinary activities. (Teacher 4)

The subject domains involved in the activities must in some sense meet and use each other properly. Subjects are not actually co-operating when the co-operation is reduced to a parasitic process where one of the subjects *de facto* is not gaining anything as described in the above quote.

MODELING INTERDISCIPLINARY ACTIVITIES INVOLVING MATHEMATICS AND PHILOSOPHY

The purpose of developing a didactical model for interdisciplinary activities involving mathematics and philosophy is, as mentioned earlier on, multiple. The model should function as a link between educational theory and the daily teaching practice in mathematics, both in the development of new activities, the carrying through of already planned ones and the evaluation of completed activities. The model gets inspiration from the work of Michelsen (2001, 2005a, 2005b), and a former version was presented at MACAS 1 and described in Iversen (2005). The didactical model consists of three phases—*the horizontal intertwining, the vertical structuring* and *the horizontal propagation.* Freudenthal (1991) introduced the idea of two different types of mathematization in an educational context—horizontal and vertical mathematization. In the horizontal mathematization students develop mathematical tools that help them organize and work with mathematical problems situated in real-life situations. The process of reorganizing the mathematical system itself Freudenthal designates vertical mathematization. Also Harel & Kaput (1991) sees a distinction between horizontal and vertical growth of mathematical knowledge. They associate the term horizontal growth with the translation of mathematical ideas between extra-mathematical situations (and models of these) and across other representation systems. By vertical growth is understood the construction of new mathematical conceptual systems.

THE HORIZONTAL INTERTWINING

As mentioned by some of the interviewed teachers interdisciplinary activities involving mathematics very often end up as fictitious constructs without much relevant mathematical content. In the first phase of a cross-curricular collaboration the attention should be centered on the importance of obtaining a real intertwining of the involved subjects. Such a curricular intertwining involves considerations about which fields of study, problems and methods in mathematics and the other subjects involved that have potentiality as interdisciplinary elements. Such elements must not originate from oversimplified lingual similarities among the subjects, but instead from considerations about how these elements can be used later in the continued learning of e.g., mathematics. This kind of intertwining of the subjects' core subject matter the students will often experience as "the meeting of different subjects," and the term of *horizontal* refers therefore to the students pre-understanding of the chosen curricular element as belonging to both mathematics and another involved subject, but not necessarily as an subject-exceeding element. Often the students do not consider ideas to be related because of their logically connection, but because they are being used together in the same kind of problem solving situations (Lesh & Doerr, 2003; Lesh & Sriraman, 2005). Michelsen et al. (2005a) suggest the term horizontal linking to describe the process of identifying contexts across mathematics and other subjects of science that are suitable for integrated modeling courses. I will here suggest the notion of *horizontal intertwining* to describe a related process of identifying and characterizing interdisciplinary problems and context suitable for integrating the subjects of mathematics and philosophy, thereby emphasizing the broader scope the integration of mathematics with a subject not from the natural sciences demands.

The interdisciplinary activities should be chosen so they set up non-routine problems, which in order to be solved properly, need the involvement of all the involved subjects. A competence approach to the subject of mathematics contains a possibility to identify such relevant subject-exceeding elements, because this approach focuses on what the students master after going through the courses, and not on concrete curricula. As argued in the theoretical section of this paper such an approach demands a broadminded view on the notion of competencies to be able to work as an educational tool.

A horizontal intertwining of the subjects designates a weaving together of the involved subjects' core subject matter by identifying non-routine problems and contexts suitable for integrating mathematics and philosophy. In order to be able to do this one needs to clarify what constitutes such core subject matters. Furthermore, such a weaving together of subjects demands a clarification of the overall purpose of the activities. The purpose must have relevance for both mathematics and the other subjects involved in order to be justified. In practice it can span a wide field of areas; from helping the cognitive growth of the individual student (e.g., in relation to concept formation), trying to strengthen the motivation for the involved subjects or even trying to create a unified view of knowledge and science in the students.

THE VERTICAL STRUCTURING

A reasonable intertwining of the involved subjects facilitates the possibility that the student can identify with the cross-curricular aspects of the chosen problems, and thereby engage meaningfully in the activities. A clarification of the overall purpose with the activities will from the beginning help the teacher to follow the students' cognitive development along the activities. Such observations will often involve that the mathematics teacher abandons the usual authoritarian role and take on a more *guide-like* function instead.[7] From a combination of the involved subjects' core subject matter the student should under suitable guidance and activity go through a cognitive development—a so-called vertical structuring—that will root the cross-curricular phenomenon concerned conceptually. It is crucial for a successful interdisciplinary engagement that the involved phenomena are central for the further learning of mathematics. If the purpose of the activities is the formation of new mathematical concepts the vertical structuring could be described as the construction of a new mathematical *concept image* (in the sense of Tall and Vinner, 1981). More theories describe how the formation of a new concept image in the student involves a qualitative change in the students perception of the specific concept. The change of perception is registered as a cognitive shift between perceiving the mathematical concept as an activity (or a process) and viewing the concept as an entity in itself, i.e., a kind of structure or object (Dubinsky 1991, Sfard 1991, Tall 1997, 2001).

In activities where the over-all purpose is to equip the students with a greater curricular perspective and overview we can describe the vertical structuring as the cognitive development of a new cross-curricular platform in the student, whereto new knowledge later can be attached to and grow from.

THE HORIZONTAL PROPAGATION

A successful vertical structuring should be evaluated in a greater perspective. The development of new significant concepts and connections based on interdisciplinary elements should be further developed in the different curricular domains of mathematics and philosophy. According to Lesh & Doerr (2003) the real challenge of the teacher is not only to introduce new ideas and concepts but also to create situations where the students need to express their current ways of thinking so this can be further tested and revised in directions of stronger development. In the case of mathematics the student should be allowed to use the newly learned knowledge in different mathematical activities and thereby apply, test and approve the specific mathematical concepts in question for the purpose of developing a more firm and generalized mathematical structure in the end. This is only possible if the original purpose with the activities is aimed at such a propagation of the new knowledge in other contexts. In other words the vertical structuring should be followed up by a horizontal propagation of the newly acquired structures in the students and thereby this knowledge can find its use in both mathematics and other involved subjects.

In this way the cross-curricular elements can work as a new basis, or context, for the student which can use it in the continued learning of mathematics furthermore in the development of new interdisciplinary connections between subjects thereby being able to overcome the crucial problems of *transfer* mentioned in the theoretical section of this paper. This is the true gain of such interdisciplinary activities.

After the carrying through of a longer cross-curricular course one of the interviewed teachers describes an example of what could be characterized as a horizontal propagation as follows

> I see the acquired competencies applied in many different places. They [the students] simply travel faster over the learning-ground. One can say that they fundamentally have a greater prerequisite for both conceptual entities and in working contexts. (Teacher 1)

DESIGNING RELEVANT ACTIVITIES INVOLVING MATHEMATICS AND PHILOSOPHY

After sketching the different components that make up the didactical model it should be illustrated how it can be used in the development of relevant interdisciplinary activities between mathematics and philosophy. Here we consider the special case of proof and proving in mathematics and philosophy.

First we need to identify relevant non-routine problems, topics or phenomena which can function as curricular-exceeding elements between the two subjects and thereby establish a reasonable horizontal intertwining. We can use a competence approach to the curriculums of mathematics and philosophy respectively, hereby focusing on what cognitive qualities the two subjects aim at developing in the students. Common to the two subjects is a (seemingly endless) search for logically healthy arguments and conclusions and the ability to follow and judge such kind of reasoning therefore belongs to the core subject matter in both mathematics and philosophy. In planning the activities we can therefore reasonably focus on developing some sort of *reasoning competence* as mentioned earlier. This involves an ability to compare and differentiate the different kinds of argumentation used by the two subjects, but also the ability to dive into specific arguments from each subject and be able to follow and judge such specific reasoning.

In all of the school's different subjects the students' ability to argue clearly and reason reasonably plays an important role, and a development of this capacity is a key area in both mathematics and philosophy. Philosophy is in fact often characterized as a subject that tries to generate and develop the students' ability to understand and use forms of argumentation and knowledge that cut across the school's different disciplines and dimensions.

Mathematical reasoning takes many forms but is in its clearest form crystallized as actual proofs. The power to give a definite proof for a certain conjecture is characteristic for the subject of mathematics and the students' knowledge about the meta-concept of *proof* is, as argued earlier in this paper, therefore central in the teaching activities in the high school. In philosophy the idea of proof also plays a key role. Earlier on, philosophers tried to transfer the mathematical (in some sense Euclidean) idea of proof to actual philosophical arguments. The most famous philosophical "proofs" are the proofs of the existence of God. These were put forward by e.g., Anselm of Canterbury and Thomas Aquinas, who both believed that giving a formal proof of the existence of God was actually possible. The high school teachers who took part in the interviews also highlighted argumentation and the concept of proof as phenomena that could transcend the gap between the subjects of mathematics and philosophy and thereby overcome the problem of transferring mathematical knowledge to other contexts and domains.

To sum up we have, starting from a wish to advance the students' ability to argue and reason within mathematics and philosophy identified *the concept of proof* as a concrete topic suitable for a curricular intertwining of the two involved subjects.

The activities originate from a study of which role the actual proving of statements and conjectures holds within the two subjects. What constitutes a proof? At what point can we say we actually have proven something? And what kind of knowledge does a proof give us? Is it true? Is it unchangeable?[8] In practice one could use simple proofs, easy for the students to master mathematically, such as small proofs from the classical *Elements* by Euclid himself (Euclid, 2002). E.g. using the proof that *the sum of the angles in a (Euclidean) triangle* is equal to the sum of two right angles or the proof of *the Pythagorean theorem*. Then comparing these to actual proofs of philosophical character e.g., a modern version of *Anselms Ontological proof* of the existence of God. It is important that the students subsequently are placed in different situations where they themselves are forced to work out small proofs thereby experiencing the process of trying to argue for a conjecture. This will enable the students to apply, test and further develop their understanding of the concept of proof. An understanding that (hopefully) in time will evolve further and be a useful tool for the students.

Through an experimenting approach, as described above, to the idea of proof a vertical structuring of the meta-concept *proof* should be developed. At the same time focus is on the students' ability to separate different kinds of argumentation. Most of the interviewed teachers agree that this would be of significant importance in the students' continued engagement with both mathematics and philosophy.

A vertical structuring of the concept of proof subsequently work as a structure which must be applied, re-valued and tested further in the daily teaching practice that follows within both subjects. Hereby obtaining a horizontal propagation of the newly acquired knowledge which results in a greater basis or context for the further learning and understanding of both mathematics and philosophy.

The Danish Ministry of Education has recently published an Education Manual for the high schools. The manual focuses on interdisciplinary activities and a large part is devoted to paradigmatic examples of concrete activities. In this manual I've contributed to more fully describe activities between mathematics and philosophy as the one sketched above.[9]

CONCLUSION

In the paper a didactical model which should function as a concept frame for the development, completion and evaluation of interdisciplinary activities involving mathematics and philosophy, was presented. The model consists of three phases that these activities involve; *The horizontal intertwining, the vertical structuring* and *the horizontal propagation*. Although the model is presented as linear, the process of going through the different phases is in some sense to be understood as an iterative process that can be run through several times by each student.

In the description of the first phase it was argued that it is of great importance that the actual mathematical content in interdisciplinary activities is not reduced to simple instrumental activities. Instead one should seek to identify and characterize interdisciplinary phenomena and contexts which can facilitate a proper intertwining of the different subjects involved by setting up relevant non-routine problems which need the involvement of both mathematics and philosophy to be answered. This can be enabled by a competence-approach as to what constitute mathematical skills. Such an approach is broader than the usual curriculum-approach to mathematics which often works as a drag to the development of successful interdisciplinary activities.

The model's second phase describes how the students' engagement in the planned activities should facilitate a vertical structuring which leads to the development of new conceptual systems, objects or contexts in the student. This can appear as a formation of new mathematical concept images, by which the interdisciplinary phenomenon considered, conceptually is anchored. This can work as a further basis in the students' continued learning of both mathematics and philosophy.

Finally the third phase focuses on how ongoing activities involving the newly acquired constructions are the overall purpose with all interdisciplinary activities. Furthermore it is argued that the cross-curricular phenomenon should be applicable in the daily teaching practice through a horizontal propagation of the considered phenomenon in both mathematics and the other subjects involved.

An anchoring of the model in the daily teaching practice was sought through a series of qualitative interviews of Danish high school teachers. Furthermore the model was illustrated through a design of a concrete interdisciplinary activity between mathematics and philosophy, and it was thereby argued how the model can be used to develop concrete interdisciplinary activities between these two subjects. The sketched activities take the concept of proofs and proving as a starting point and centers themselves around argumentation and reasoning in both mathematics and philosophy.

As the modeling of such activities is still (and perhaps always) a work-in-progress the presented model is somewhat tentative in its nature. The model originates from a wish to develop a concept frame for interdisciplinary activities between mathematics and philosophy, and found inspiration in the work of Michelsen (2001, 2005a, 2005b) which centers about interdisciplinary activities between mathematics and physics. A further perspective is to continue the work of developing concrete teaching activities, as well as trying to adapt and evaluate the model's strengths and weaknesses as a didactical tool to integrating the subjects of mathematics and philosophy. The author, therefore, invites all interested readers to further test and revise the model as well as concrete realizations and afterwards sharing experiences which hopefully will lead to the improvement of the didactical model as a result.

REFERENCES

Dubinsky, E. (1991). Reflective abstraction in advanced mathematical thinking. In D. Tall (Ed.) *Advanced Mathematical Thinking* (pp. 95–126), Kluwer Academic Publishers.

Dahland, G. (1998). *Matematikundervisning I 1990-talets gymnasieskola. Ett studium av hur didaktisk tradition har påverkats av informationsteknologins verktyg,* Institution för pedagogik, Göteborgs universitet, Göteborg.

Dreyfus, T. (1999). Why Johnny can´t prove, *Educational Studies in Mathematics, 38,* 85–109.

Euclid, (Densmore, D. and Heath, T.L.) (2002): *Euclid's Elements,* Green Lion Press.

Freudenthal, H. (1991). *Revisiting Mathematics Education: China Lectures.* Kluwer Academic Publishers.

Gravemeijer, K. (1997): Instructional design for reform in mathematics education. In Beishuizen, Gravemeijer & van Lishout (Eds.): *The Role of Contexts and Models in the Development of Mathematical Strategies and Procedures,* Utrecht, CD β Press, 13–34.

Gräber, W. *et al.* (2001). Scientific literacy: From theory to practice, In Behrendt, H. et al. (Eds.), *Research in Science Education—Past, Present, and Future* (pp. 61–70), Kluwer Academic Publishers.

Harel, G. & Kaput, J. (1991). The role of conceptual entities and their symbols in building advanced mathematical concepts, In D. Tall, (Ed.), *Advanced Mathematical Thinking* (pp. 82–94), Kluwer Academic Publishers.

Hatano, G (1996). A conception of knowledge acquisition and its implications for mathematics Education. In L. Steffe et al., (Eds), *Theories of Mathematical Learning* (pp. 197–217), Hillsdale, NJ: Erlbaum.

Hazzan, O. & Zazkis, R. (2005). Reducing Abstraction: The Case of School mathematics. *Educational Studies in Mathematics,* 101–119.

Iversen, S. M. (2005): Building a model for cross-curricular activities between mathematics and Philosophy. In A. Beckmann, C. Michelsen, & B. Sriraman (Eds.), *Proceedings of the 1st International Symposium of Mathematics and its Connections to the Arts and Sciences* (pp. 142–151). The University of Education, Schwäbisch Gmünd, Franzbecker Verlag.

Lesh, R. & Doerr, H. M. (2003): *Beyond Constructivism – Models and Modeling Perspectives on Mathematics Problem Solving; Learning, and Teaching,* Lawrence Erlbaum Associates.

Lesh, R. & Sriraman, B. (2005). John Dewey Revisited- Pragmatism and the models-modeling perspective on mathematical learning. In A. Beckmann et al (Eds.), *Proceedings of the 1st International Symposium on Mathematics and its Connections to the Arts and Sciences.* May 18–21, 2005, University of Schwaebisch Gmuend: Germany.Franzbecker Verlag, pp. 32–51.

Michelsen, C. (2001): *Begrebsdannelse ved domæneudvidelse–Elevers tilegnelse af funktionsbegrebet i et integreret undervisningsforløb mellem matematik og fysik,* Ph.D. dissertation, University of Southern Denmark.

Michelsen, C., Glargaard, N. & Dejgaard, J. (2005a): Interdisciplinary competences—integrating mathematics and subjects of natural sciences. In M. Anaya, C. Michelsen (Editors), *Proceedings of the Topic Study Group 21: Relations between mathematics and others subjects of art and science. The 10th International Congress of Mathematics Education,* Copenhagen, Denmark

Michelsen, C. (2005b): Expanding the Domain—Variables and functions in an interdisciplinary context between mathematics and physics. In A. Beckmann, C. Michelsen, & B. Sriraman (Eds.)., (2005). *Proceedings of the 1st International Symposium of Mathematics and its Connections to the Arts and Sciences.* The University of Education, Schwäbisch Gmünd, Franzbecker Verlag, pp. 201–214.

Mitchell, M. (1993). Situational interest: its multifaceted structure in the secondary school mathematics classroom, *Journal of Educational Psychology, 85,* 424–436.

Niss, M. (1999). Aspects of the nature and state of research in mathematics education, *Educational Studies in Mathematics, 40,* 1–24.

Niss, M. & Jensen, T. H. (Eds.) (2002). *Kompetencer og matematiklæring. Ideer og inspiration til udvikling af matematikundervisningen i Danmark*, The Danish Ministry of Education, Copenhagen.

Sfard, A. (1991). On the Dual Nature of Mathematical Conceptions: Reflections on Processes and Objects as Different Sides of the Same Coin, *Educational Studies in mathematics, 22*, 1–36.

Sriraman,B. (2004). Re-creating the Renaissance. In M. Anaya, C. Michelsen (Editors), *Proceedings of the Topic Study Group 21: Relations between mathematics and others subjects of art and science*: *The 10th International Congress of Mathematics Education*, Copenhagen, Denmark, pp. 14–19

Tall, D. & Vinner, S. (1981): Concept Image and Concept Definition in mathematics with particular reference to limits and continuity, *Educational Studies in Mathematics, 12*, 151–169.

Tall, D. et al. (1997): *What is the object of the encapsulation of a process?*, *Proceedings of MERGA*, Rotarua, *2*, 132–139.

Tall, D. & Gray, E. (2001): Relationships between embodied objects and symbolic procepts: An explanatory theory of success and failure in mathematics, *Proceedings of PME25*, Utrecht, pp. 65–72.

Vinner, S. (1991): The role of definitions in the teaching and learning of mathematics. In D. Tall, (Ed.), *Advanced Mathematical Thinking*, Kluwer Academic Publishers, pp. 65–81.

NOTES

1. This follows Sriraman (2004) who argued that students are used to viewing knowledge through the discrete lens of disjoint school subjects.

2. The symposium took place 18-21 May 2005 in Schwäbish Gmünd, Germany. See Beckmann, A., Michelsen, C., & Sriraman, B (Eds.)., (2005). *Proceedings of the 1st International Symposium of Mathematics and its Connections to the Arts and Sciences*. The University of Education, Schwäbisch Gmünd, Germany, Franzbecker Verlag.

3. Related should here be understood in a common way. The subject of mathematics is e.g. is supposed to be more *related* to physics than to English.

4. My own translation from Danish (ibid.).

5. The fact that some of the asked questions particularly involved a reference to the Danish high schools(as opposed to any high schools) was because I wanted to find out which effect a forthcoming reform of the Danish high schools would have on the daily teaching practice. Most of questions asked involved only general educational components, and did not hold any particular reference to any Danish conditions.

6. All the interviews were conducted in Danish, and the quotes given in the text is therefore my own translation. The text in the brackets is my insertions. They are there to give the right coherence in the teachers statements. The interviewed teachers are here given only a number, but all the quotes given in this paper are approved by the particularly teacher concerned.

7. For a more developed description of this shift in the teachers role in the class-room, see e.g., Gravemeijer (1997).
8. All questions Niss (1999) emphasized as extremely difficult for students to answer properly.
9. The manual can be found at http:us.uvm.dk/gymnasie/vejl/?menuid=15 (unfortunately only in Danish).

ACKNOWLEDGEMENT

Reprint of Iversen, S. (2006). Modelling interdisciplinary activities involving mathematics and philosophy. *The Montana Mathematics Enthusiast, 3*(1), 85–98. © The Montana Mathematics Enthusiast.

CHAPTER 12

INTEGRATING ENGINEERING EDUCATION WITHIN THE ELEMENTARY AND MIDDLE SCHOOL MATHEMATICS CURRICULUM

Lyn D. English
Queensland University of Technology

Nicholas G. Mousoulides
Cyprus University of Technology

Many nations are experiencing a decline in the number of graduating engineers, an overall poor preparedness for engineering studies in tertiary institutions, and a lack of diversity in the field. Given the increasing importance of mathematics, science, engineering, and technology in our world, it is imperative that we foster an interest and drive to participate in engineering from an early age. This discussion paper argues for the integration of engineering education within the elementary and middle school mathematics curricula. In doing so, we offer a definition of engineering education and address its core goals; consider some perceptions of engineering and engineering education held by teachers and students; and offer one approach to promoting engineering education within the elementary and middle school mathematics curriculum, namely through mathematical modeling.

Interdisciplinarity, Creativity, and Learning, pages 165–175
Copyright © 2009 by Information Age Publishing

INTRODUCTION

The world's demand for skilled workers in mathematics, science, engineering, and technology is increasing rapidly yet supply is declining across several nations, with the number of engineers graduating from U.S. institutions slipping 20 percent in recent years (National Academy of Sciences, 2007; OECD, 2006). To complicate matters, recent data reveal waning student interest in engineering, poor educational preparedness, a lack of diverse representation, and low persistence of current and future engineering students (Dawes & Rasmussen, 2007; Lambert, Diefes-Dux, Beck, Duncan, Oware, & Nemeth, 2007). We need young scholars to be involved in the next generation of innovative ideas that support our society's needs. This interest and drive to participate in engineering must be fostered from an early age. To date, there has been very limited research on integrating engineering experiences in the elementary and middle school curricula.

The field of engineering education is just emerging, with a number of questions identified for attention. These include: "What constitutes engineering thinking for elementary/middle school children?" "How can the nature of engineering and engineering practice be made visible to young learners?" "How can we integrate engineering experiences within existing school curricula?" "What engineering contexts are meaningful, engaging, and inspiring to young learners?" and "What teacher professional development opportunities and supports are needed to facilitate teaching engineering thinking within the curriculum?" (Cunningham & Hester, 2007; Dawes & Ramussen, 2007; Kuehner & Mauch, 2006; Lambert et al., 2007).

This paper begins a discussion on some of the above issues. In particular, we offer a definition of engineering education and address its core goals; we consider some perceptions of engineering and engineering education held by teachers and students; and we offer one approach to promoting engineering education within the elementary/middle school mathematics curriculum.

WHAT IS ENGINEERING EDUCATION?

The field of engineering has been described in many ways. Wulf (1998) referred to engineering as "design under constraint. Engineering is synthetic–it is creating, designing what can be, but it is constrained by nature, by cost, by concerns of safety, reliability, environmental impact, and many other..." (p. 1). Given the urgent need for more engineers in diverse fields, it is timely to address the introduction of engineering-based experiences (engineering education) within the elementary and middle school curricula.

Engineering education in the elementary/middle school is a significant, nascent field of research that aims to foster children's appreciation and understanding of what engineers do and how engineering shapes the world around them. Engineering-based experiences encourage students to generate effective tools for dealing with our increasingly complex, dynamic, and powerful systems of information (Zawojewski, Hjalmarson, Bowman, & Lesh, 2008). The experiences build on children's curiosity about the scientific world, how it functions, and how we interact with the environment, as well as on children's intrinsic interest in designing, building, and dismantling objects in learning how they work (Petroski, 2003). Integrating engineering experiences within the elementary/middle school curricula also can help children appreciate how their learning in these domains can apply to the solution of real-world problems.

One of the foremost institutions that are introducing engineering into the elementary mathematics and science curricula is the National Center for Technological Literacy at the Museum of Science in Boston (Cunningham & Hester, 2007). The goals and activities of their *Engineering is Elementary* program are well suited for integration within the school curricula and provide fertile ground for interdisciplinary research. Cunningham and Hester (2007) have identified three core goals of their *Engineering is Elementary* program, namely, to: (a) Increase children's technological literacy; (b) Increase elementary educators' abilities to teach engineering and technology to their students; and (c) Modify systems of education to include engineering at the elementary level.

With respect to the first goal, it is important that children develop: a knowledge of what technology and engineering in its various forms entail and what engineers do; an appreciation of how engineering and technology have shaped so many facets of our world, and how society influences and is influenced by engineering and technology; and an understanding and appreciation of how mathematics and science are applied to solving engineering problems, of which there are multiple solutions. A facility in applying an engineering design process in solving real-world problems is also an important component of this first goal, as we address in a later section.

With respect to the second goal, for teachers to be able to effectively integrate engineering experiences within the primary mathematics and science curricula, they need professional development and appropriate resources to scaffold their understanding and pedagogical strategies. As we indicate next, the vast majority of elementary school teachers have had little or no education about engineering concepts or ways of implementing engineering experiences within the curriculum.

The third goal listed above requires targeting state education systems to provide opportunities for engineering experiences in the elementary school. Some nations are beginning to address this issue. For example, the

INSPIRE (Institute for P–12 Engineering Research and Learning) team at Purdue University is initiating and leading an advocacy effort at state and national levels to inform and influence policy-making that will increase the U.S. commitment to P–12 engineering education. In instigating these broad-scale efforts, it is imperative that teachers' and students' perceptions of engineers and engineering be addressed (Lambert et al., 2007).

PERCEPTIONS OF ENGINEERING AND ENGINEERING EDUCATION

As our society becomes increasingly dependent on engineering in its various forms, it is more important than ever that citizens have a basic understanding of what engineering entails, of what engineers do, and of the uses and implications of the technologies that they generate (Cunningham & Hester, 2007; Knight & Cunningham, 2004). However, few studies have probed students' and teachers' understanding of these issues. The little research that has been conducted has indicated that people generally do not understand what engineers do, despite being surrounded by the products of engineering in their everyday world (e.g., Cunningham, Lachapelle, & Lindgren-Streicher, 2005; Davis & Gibbon, 2002; Knight & Cunningham, 2004; Lambert et al., 2007). For example, Cunningham et al. (2005) found that school teachers are more likely to believe that engineers build rather than supervise the construction of buildings.

Little appreciation of the role of mathematics, science, and technology in engineering was noted by Lambert et al. (2007) in their study of P–6 teachers' perceptions and understanding of engineering. Furthermore, few teachers offered ideas about the need for team work, communication, and global/societal perspectives in engineering. Nor did the teachers identify engineering as creative but bounded by constraints.

Teachers' limited views on accessibility to engineering courses was noted by Douglas, Iversen, and Kalyandurg (2004). Their study indicated that teachers generally believe that engineering has a major impact on their daily lives and that implementing engineering concepts within the curriculum is certainly warranted. However, there is the belief that engineering is not an option for a large number of students and that the field is very difficult to enter at the university level.

Findings from the scant studies that have explored school students' conceptions of engineering indicate that students generally do not understand what engineers do. For example, Cunningham et al. (2005) administered their *What is an Engineer?* instrument to over 6000 elementary school students and found that they strongly conflate construction workers and auto mechanics with engineers. This is perhaps not surprising, given that these fields are not traditionally populated by women, suggesting that such narrow views of engineering might be one reason for the lower number of females who enter engineering courses. It is imperative that we help our students and teachers understand the range and type of work that engineers do and appreciate the role of engineering in advancing society.

PROMOTING ENGINEERING EDUCATION WITHIN THE ELEMENTARY MATHEMATICS CURRICULUM

Engineering-based problem experiences readily align themselves with those of the mathematics (and science) curriculum. For example, engaging children in hands-on, real-world engineering experiences involves them in design process cycles that utilize powerful mathematical problem solving and reasoning processes. Many such design process cycles exist. The *Engineering is Elementary* program (Cunningham & Hester, 2007) emphasizes a process cycle involving the components: ask, imagine, plan, create, and improve (see Figure 12.1). The design process can begin at any component, with movement back and forth between the components occurring numerous times.

We address here one means to designing and implementing engineering experiences within the mathematics curriculum, one that utilizes a comprehensive variation of the above design process cycle, namely, a models and modeling approach (Diefes-Dux & Duncan, 2007; English, 2007, in press; Lesh & Doerr, 2003). In adopting this approach, we present real-world engineering situations in which children repeatedly express, test, and refine or revise their current ways of thinking as they endeavour to create a struc-

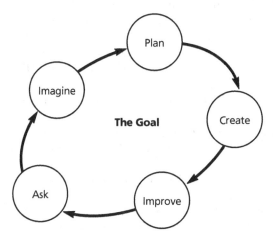

Figure 12.1   A cyclic process of engineering design.

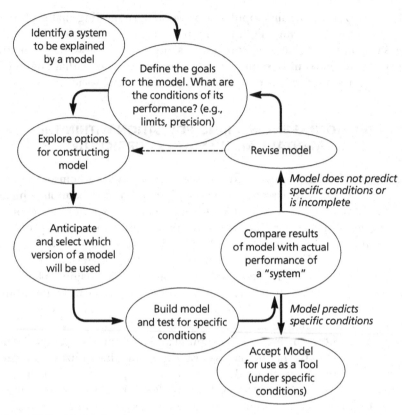

Figure 12.2   Model development process.

turally significant product—namely, a model that can be used to interpret, explain, and predict the behaviour of one or more systems defined by the problem. Diefes-Dux, Osburn, Capoobianco, and Wood (2008) describe the development of such models in terms of four key, iterative activities, which they represent in a flow diagram shown in Figure 12.2.

1. Understanding the context of the problem and the system to be modeled;
2. Expressing/testing/revising a working model;
3. Evaluating the model under conditions of its intended application; and
4. Documenting the model throughout the development process.

Students engage in these iterative activities when they undertake model-eliciting engineering-based problems, which we address next.

MODEL-ELICITING ENGINEERING-BASED PROBLEMS

The engineering-based problems we have implemented in elementary and middle-school classrooms are realistic open-ended problems where a client requires a team of workers to generate a product (a model) for solving the given problem. The model identifies a *process* that the client can use to solve the problem. We provide an example here of an environmental engineering problem, namely, the *Water Shortage Problem* (Mousoulides, 2008; we present part of the problem in the appendix; space prevents the entire problem being included but this can be obtained from the authors).

In the *Water Shortage Problem* students are sent a letter from a client, the Ministry of Transportation, who needs a means of (model for) selecting a country that can supply Cyprus with water during the next summer period. The letter asks students to develop a procedure/model using the data provided from three countries and also to obtain extra data, using available tools such as Google Earth, maps, and the Web. The quantitative and qualitative data provided for each country include water supply per week, water price, tanker capacity, and ports' facilities. Students can also obtain data about distance between countries, major ports in each country, and oil consumption. After students have developed their model, they write a letter to the client detailing how their model selects the best country for supplying water. Upon completion of their letter, students receive a second letter from the client including data for two more countries; students are asked to test their model on the expanded data and improve their model, if needed.

Engineering problems such as the *Water Shortage Problem* are designed to be thought revealing. The development of engineering problems is guided by six principles for designing model eliciting activities (Diefes-Dux, Hjalmarson, Miller, & Lesh, 2008). A brief description of the principles and how these have been applied in the design of the Water Shortage Problem are presented below.

According to the *Model Construction Principle*, the design of the problem requires the creation of a model by the student team; the model is often a procedure for carrying out a task. The Water Shortage Problem requires students to develop a procedure for selecting the best among different countries that can supply Cyprus with water, taking into consideration both qualitative and quantitative data as well as other necessary data students should obtain.

The Water Shortage Problem is inline with the *Reality Principle*, which requires the problem to be situated in an authentic engineering context. The Water Shortage Problem is an authentic environmental engineering problem that is also of great interest for the majority of students. According to the third principle, the *Self-Assessment Principle*, the design of the problem provides opportunities for students to work in their teams to assess the appropriateness of their models for selecting the best country. The problem also requires students to produce a documentation of their model (writing a letter to the Ministry of Transportation), which meets the *Model Documentation Principle* requirement. The problem also takes into account the *Model Shareability and Reusability* and *Effective Prototype Principles*—the models that students develop should be applicable to structurally similar engineering problems.

Engineering problems that meet the above principles, such as the *Water Shortage Problem*, are designed so that multiple solutions of varying mathematical sophistication are possible and children with a range of personal experiences and knowledge can tackle them. The products children create are documented, shareable, reusable, and modifiable mathematical models that provide teachers with a window into their students' conceptual strengths and weaknesses.

Another important feature of these problems is the opportunities provided for multiple feedback points to encourage children to rethink their models (e.g., through "what-if" questioning) and for discussion on the strengths and weaknesses of their models (with respect to the client's criteria for success) across a classroom full of alternative models. Furthermore, these modeling problems build communication (oral and written) and team-work skills, both of which are essential to success in mathematics and engineering.

CONCLUDING POINTS

We have argued here for the integration of engineering education within the elementary and middle school mathematics curriculum and have offered one approach to achieving this through mathematical modeling. Engineering-based modeling experiences provide opportunities for students to deal with multidisciplinary contexts, to identify, formulate, and solve real-world engineering problems, and to communicate their ideas effectively to others. Engineering education for younger students is a new and much-needed field of research. The elementary/middle school curriculum provides ideal opportunities for introducing children to foundational engineering ideas and principles. We consider it imperative that young scholars develop a strong curiosity and drive to learn how engineering shapes their world and supports so many of our society's needs.

REFERENCES

Cunningham, C. M., & Hester, K. (2007). Engineering is elementary: An engineering and technology curriculum for children. In *Proceedings of the 2007 American Society for Engineering Education Annual Conference & Exposition*. Honolulu, Hawaii: American Society for Engineering Education.

Cunningham, C. M., Lachapelle, C. P., & Lindgren-Streicher, A. (2005). Assessing elementary school students' conceptions of engineering and technology. In *Proceedings of the 2005 American Society for Engineering Education Annual Conference & Exposition*. American Society for Engineering Education.

Davis, L. A., & Gibbon, R. D. (2002). *Raising public awareness of engineering*. National Academy of Engineers: Washington DC.

Dawes, L., & Rasmussen, G. (2007). Activity and engagement—keys in connecting engineering with secondary school students. *Australasian Journal of Engineering Education, 13*(1), 13–20.

Diefes-Dux, H.A., & Duncan, D. (2007). Adapting Engineering is Elementary Professional Development to Encourage Open-Ended Mathematical Modeling. *Committee on K–12 Engineering Education, National Academy of Engineering, National Research Council, Workshop and Third Meeting*, Oct. 22, 2007, Keck Center of the National Academies, Engineering Education in Grades K–5.

Diefes-Dux, H.A., Hjalmarson, M., Miller, T., & Lesh, R. (2008). Model Eliciting Activities for Engineering Education. In J. Zawojewski, H.A. Diefes-Dux & K. Bowman (Eds.) *Models and Modeling in Engineering Education: Designing Experiences for All Students* (pp. 17–35). Sense Publishers.

Douglas, E., Iversen, C., & Kalyandurg, S. (2004). *Engineering in the K–12 classroom— An analysis of current practices and guidelines for the future*. ASEE Engineering K12 Center.

English, L. D. (2007). Mathematical modelling: Linking Mathematics, Science, and the Arts in the Elementary Curriculum. In A. Beckmann, C. Michelsen, & B. Sriraman (Eds.) *Proceedings of the Second International Symposium of Mathematics and its Connections to the Arts and Sciences* (pp. 5–35). University of Southern Denmark.

English, L. D. (In press). Promoting interdisciplinarity through mathematical modelling. *ZDM: The International Journal on Mathematics Education, 41*(1).

Knight, M., & Cunningham, C. (2004). Draw an engineer test (DAET): Development of a tool to investigate students' ideas about engineers' and engineering. In *Proceedings of the 2004 American Society for Engineering Education Annual Conference & Exposition*. Salt Lake City, Utah: American Society for Engineering Education.

Kuehner, J. P., & Mauch, E. K. (2006). Engineering applications for demonstrating mathematical problem-solving methods at the secondary education level. *Teaching Mathematics and its Applications, 25*(4), 189–195.

Lambert, M. Diefes-Dux, H., Beck, M., Duncan, D., Oware, E., & Nemeth, R. (2007). What is engineering? – An exploration of P–6 grade teachers' perspectives. In

Proceedings of the 37th ASEE/IEEE Frontiers in Education Conference. Milwaukee, Wisconsin.

Lesh, R., & Doerr, H. M. (Eds.). (2003). *Beyond constructivism: Models and modeling perspectives on mathematic problem solving, learning and teaching.* Mahwah, NJ: LEA.

Lesh, R., Cramer, K., Doerr, H. M., Post, T., & Zawojewski, J. S. (2003). Model development sequences. In R. Lesh & H. M. Doerr, (Eds.). (2003). *Beyond constructivism: Models and modeling perspectives on mathematic problem solving, learning and teaching* (pp. 35–58). Mahwah, NJ: Lawrence Erlbaum.

Mousoulides, N. (2008). *Water Shortage Problem in Cyprus.* Unpublished Modelling Problem. University of Cyprus.

National Academy of Sciences (2007). *Rising above the gathering Storm: Energizing and employing America for a brighter economic future.* Washington, DC: National Academics Press.

Organisation for Economic Co-operation and development. (2006). *OECD Science, technology, and industry outlook 2006* (highlights). Retrieved 11th of December, www.oecd.org/dataoecd/39/19/37685541.pdf

Petroski, H. (2003). Early education. *American Scientist, 91,* 206–209.

Wulf, W. A. (1998). The Urgency of Engineering Education Reform. *The Bridge, 28*(1), 1–8.

Zawojewski, J., Hjalmarson, M., Bowman, K., & Lesh R. (2008). A modeling perspective on learning and teaching in engineering education. In J. Zawojewski, H.A. Dietes-Dux & K. Bowman (Eds.) *Models and Modeling in Engineering Education: Designing Experiences for All Students* (pp. 1–15). Sense Publishers.

APPENDIX

Water Shortage Problem

Cyprus will buy Water from Nearby Countries

Background Information

One of the biggest problems that Cyprus faces nowadays is the water shortage problem. Instead of constructing new desalination plants, local authorities decided to use oil tankers for importing water from other countries. Lebanon, Greece and Egypt expressed their willingness to supply Cyprus with water. Local authorities have received information about the water price, how much water they can supply Cyprus with during summer, tanker oil cost and the port facilities.

Problem

The local authorities need to decide from which country Cyprus will import water for the next summer period. Using the information provided, assist the local authorities in making the best possible choice. Write a letter explaining the method you used to make your decision so that they can use your method for selecting the best available option (The following table was supplied).

Country	Water Supply per week (metric tons)	Water Price (metric ton)	Tanker Capacity	Oil cost per 100 km	Port Facilities for Tankers
Egypt	3,000,000	€ 3,5	30,000	€ 20,000	Good
Greece	4,000,000	€ 2	50 000	€ 25,000	Very Good
Lebanon	2,000,000	€ 4	50 000	€ 25,000	Average

CHAPTER 13

MATHEMATICAL MODELLING IN THE EARLY SCHOOL YEARS

Lyn D. English and James J. Watters
Queensland University of Technology

ABSTRACT

In this article we explore young children's development of mathematical knowledge and reasoning processes as they worked two modelling problems (the *Butter Beans Problem* and the *Airplane Problem*). The problems involve authentic situations that need to be interpreted and described in mathematical ways. Both problems include tables of data, together with background information containing specific criteria to be considered in the solution process. Four classes of third-graders (8 years of age) and their teachers participated in the 6-month program, which included preparatory modelling activities along with professional development for the teachers. In discussing our findings we address: (a) Ways in which the children applied their informal, personal knowledge to the problems; (b) How the children interpreted the tables of data, including difficulties they experienced; (c) How the children operated on the data, including aggregating and comparing data, and looking for trends and patterns; (d) How the children developed important mathematical ideas; and (e) Ways in which the children represented their mathematical understandings.

Interdisciplinarity, Creativity, and Learning, pages 177–201
Copyright © 2009 by Information Age Publishing

INTRODUCTION

> Making modelling, generalization, and justification an explicit focus of instruction can help to make big ideas available to all students at all ages. (Carpenter & Romberg, 2004, p. 5)

We face a world that is shaped by increasingly complex, dynamic, and powerful systems of information, such as sophisticated buying, leasing, and loan plans that appear regularly in the media. Being able to interpret and work with such systems involves important mathematical processes that have been under-emphasized in many mathematics curricula. Processes such as constructing, explaining, justifying, predicting, conjecturing, and representing, as well as quantifying, coordinating, and organising data are becoming all the more important for all citizens. Mathematical modelling, which traditionally has been the domain of the secondary school years, provides rich opportunities for students to develop these important processes.

A model may be defined as "a system of conceptual frameworks used to construct, interpret, and mathematically describe a situation" (Richardson, 2004, p. viii). By engaging in mathematical modelling students identify the underlying mathematical structure of complex phenomena. Because mathematical models focus on structural characteristics of phenomena (e.g., patterns, interactions, and relationships among elements) rather than surface features (e.g., biological, physical or artistic attributes), they are powerful tools in predicting the behaviour of complex systems (Lesh & Harel, 2003). As such, mathematical modelling is foundational to modern scientific research, such as biotechnology, aeronautical engineering, and informatics (e.g., Gainsburg, 2004, submitted).

Many nations are expressing concern over the lack of their students' participation in mathematics and science (e.g., O'Connor, White, Greenwood, & Mousley 2001; U.S. President's "No Child Left Behind," 2000; http:/www.ed.gov/inits/nclb/titlepage.html). However, research has shown that low levels of participation and performance in mathematics are not due primarily to a lack of ability or potential, but rather, to educational practices that deny access to meaningful high-quality learning experiences (e.g., Tate & Rousseau, 2002). Many of these under-achieving students show exceptional abilities to deal with sophisticated mathematical constructs when these understandings are grounded in their personal experiences and are expressed in familiar modes of representation and discourse (Lesh, 1998). It has been shown that a broader range of students emerge as being highly capable, irrespective of their age or classroom mathematics achievement level when they participate in mathematical modelling experiences (Doerr & English, 2003; Lesh & Doerr, 2003; Lamon, 2003). As a consequence, improvements in students' confidence in, and attitudes towards, mathematics and mathematical problem solving become evident.

The primary school is the educational environment where all children should begin a meaningful development of mathematical modelling (Carpenter & Romberg, 2004; Jones, Langrall, Thornton, & Nisbet, 2002, Lehrer & Schauble, 2003; NCTM, 2000). However, as Jones et al. note, even the major periods of reform and enlightenment in primary mathematics do not seem to have given most children access to the deep ideas and key processes that lead to success beyond school.

The study reported here sought to redress this situation by engaging young children and their teachers in a 6-month program, which included preparatory modelling activities culminating in two modelling problems. This paper explores the children's development of mathematical knowledge and reasoning processes as they worked the two modelling problems over several weeks.

MATHEMATICAL MODELLING FOR YOUNG LEARNERS

Until recently, mathematical modelling (of the type addressed here) has not been considered within the early school curriculum; Rather, it has been the domain of the secondary year levels (e.g., Stillman, 1998). We argue that the rudiments of mathematical modelling can and should begin in the primary school where young children already have the basic competencies on which modelling can be developed (Carpenter & Romberg, 2004; Diezmann, Watters, & English, 2002; Lehrer & Schauble, 2003; NCTM, 2000; Perry & Dockett, 2002). Indeed, as Carpenter and Romberg documented recently,

> Our research has shown that children can learn to model, generalize, and justify at earlier ages than traditionally believed possible, and that engaging in these practices provides students with early access to scientific and mathematical reasoning. Until recently, however, these practices have not been much in evidence in the school curriculum until high school, if at all. (p. 4)

Mathematical modelling activities differ from the usual problems that young children meet in class. Problem solving in the early years has usually been limited to examples in which children apply a known procedure or follow a clearly defined pathway. The "givens," the goal, and the "legal" solution steps are usually specified unambiguously—that is, they can be interpreted in one and only one way. This means that the interpretation process for the child has been minimalised or eliminated. The difficulty for the child is basically working out how to get from the given state to the goal state. Although not denying the importance of these existing problem experiences, it is questionable whether they address adequately the mathematical knowledge, processes, representational fluency, and social skills that our children need for the 21st century (English, 2002; Carpenter & Romberg, 2004; Steen, 2001).

In contrast to the typical "word problems" presented to young children, mathematical modelling problems involve authentic situations that need to be interpreted and described in mathematical ways (Lesh & Harel, 2003). The information given, including the goal itself, can be incomplete, ambiguous, or undefined (as often happens in real life). Furthermore, information contained in these modelling problems is often presented in representational form, such as tables of data or visual representations, which must be interpreted by the child.

In recent years there has been a strong emphasis on providing young children with equal access to powerful mathematical ideas (Carpenter & Romberg, 2004; Diezmann & Watters, 2003; English, 2002; Perry & Dockett, 2002). Mathematical modelling problems provide one avenue for meeting this challenge. Key mathematical constructs are embedded within the problem context and are elicited by the children as they work the problem. The generative nature of these problems means that children can access mathematical ideas at varying levels of sophistication. For example, as we indicate later, young children can access informal ideas of rate by considering how time and distance could determine the winner of a paper plane contest.

The importance of argumentation in young children's mathematical development has also been highlighted in recent years (e.g., Perry & Dockett, 2002; Yackel & Cobb, 1996). Although Piaget (e.g., Inhelder & Piaget, 1958) claimed that the ability to argue logically is beyond the realms of young children, recent work has demonstrated otherwise (e.g., Dockett & Perry, 2001). As Perry and Dockett (2002) noted, it is important for us to be aware of and nurture the early genesis of argumentation, especially since it will form the basis of mathematical proof in later years. Mathematical modelling activities provide a solid basis for young children's development of argumentation because they are inherently social experiences (Zawojewski, Lesh, & English, 2003) and foster effective communication, teamwork, and reflection. The modelling activities are specifically designed for small-group work, where children are required to develop sharable products that involve descriptions, explanations, justifications, and mathematical representations. Numerous questions, conjectures, conflicts, resolutions, and revisions normally arise as children develop, assess, and prepare to communicate their products. Because the products are to be shared with and used by others, they must hold up under the scrutiny of the team members.

DESCRIPTION OF THE STUDY

Setting and Participants

All four third-grade classes (8 years-old) and their teachers from a state school situated in a middle-class suburb of Brisbane, participated in the study. The principal and assistant principal provided strong support for the project and attended some of the workshops and debriefing meetings that we conducted with the teachers.

Tasks

In collaboration with the teachers, we developed four preparatory activities, which were followed by two modelling problems.

The Preparatory Activities. These were designed to develop children's skills in: (a) interpreting mathematical and scientific information presented in text and diagrammatic form; (b) reading simple tables of data; (c) collecting, analysing, and representing data; (d) preparing written reports from data analysis; (e) working collaboratively in group situations, and (f) sharing end products with class peers by means of verbal and written reports. For example, one preparatory activity involving the study of animals required the students to read written text on "The Lifestyle of our Bilby," which included tables of data displaying the size, tail length, and weight of the two types of Bilbies. The children answered questions about the text and the tables.

The modelling problems. The contexts of the modelling and preparatory activities were chosen to fit in with the teachers' classroom themes, which included a study of food, animals, and flight. The first modelling activity, "Farmer Sprout," comprised a story about the various types of beans a farmer grew, along with data about various conditions for their growth. After responding to questions about the text, the children were presented with the "Butter Beans" problem comprising two parts. The children had to examine two tables of data displaying the weight of butter beans after 6, 8, and 10 weeks of growth under two conditions (sunlight and shade; see Table 13.1).

Using the data of Table 13.1, the children had to (a) determine which of the conditions was better for growing butter beans to produce the greatest crop. As a culminating task the children were required to write a group letter to Farmer Sprout in which they outlined their recommendation and explained how they arrived at their decision; and then (b) predict the weight of butter beans produced on week 12 for each type of condition. The children were to explain how they made their prediction so that the farmer

TABLE 13.1 Data presented for the Beans Problem

	Sunlight				Shade		
Butter Bean Plants	Week			Butter Bean Plants	Week		
	6	8	10		6	8	10
Row 1	9 kg	12 kg	13 kg	Row 1	5 kg	9 kg	15 kg
Row 2	8 kg	11 kg	14 kg	Row 2	5 kg	8 kg	14 kg
Row 3	9 kg	14 kg	18 kg	Row 3	6 kg	9 kg	12 kg
Row 4	10 kg	11 kg	17 kg	Row 4	6 kg	10 kg	13 kg

could use their method for other similar situations. On completion of the activity, each group reported back to the class. Following the reporting back, the group's peers asked questions and provided constructive feedback.

The second modelling activity, "The Annual Paper Airplane Contest," (see the Appendix) presented children with a newspaper article that described an annual airplane contest involving the flight performance of paper airplanes. The children were given information regarding the construction of the planes and the rules for the flight contest. After completing a number of comprehension questions, the children were given the problem information and associated investigation shown in the Appendix.

Procedures

Teacher meetings. We implemented a number of workshops and debriefing sessions for the teachers throughout the year. We conducted two half-day workshops with the teachers in term 1 to introduce them to the activities and to plan the year's program more thoroughly. In these workshops, the teachers worked on the activities they were to implement and identified various approaches to solution. Two more workshops were conducted during the middle and at the end of the year for planning and reflective analysis of the children's and teachers' progress. Several shorter meetings were also conducted throughout the year, including those before and after the teachers had implemented each activity. During these debriefing sessions the teachers discussed with the researchers issues related to student learning, the activities, and implementation strategies.

Task implementation. The preparatory activities were implemented weekly by the teachers towards the end of first term and part of second term. During the remainder of second term and for all of third term, the teachers implemented, on a weekly basis, the two modelling problems. There was approximately one month's lapse between children's completion of the Butter Beans Problem and the Airplane Problem.

Each modelling activity was explored over 4–5 sessions of 40 minutes duration each and conducted as part of the normal teaching program. After an initial whole class introduction to the modelling activity, the children worked independently in groups of 3 to 4 on the activity. The teachers monitored each group and provided scaffolding where necessary. Such scaffolding included questioning children for explanation and justification, challenging the children, querying an inappropriate action, and providing overall encouragement and motivation. The teachers also focussed on supporting children's writing and the development of group skills. In the final session the students provided a group report to the class and their conclusions were discussed.

Each of the teachers had previously established procedures for group work and for class reporting. For example, each group of children had a group-appointed manager who was responsible for organising materials and keeping the group on task. The importance of sharing ideas as well as explaining answers was also emphasised in class group work.

DATA COLLECTION AND ANALYSIS

In each of the four classes, we videotaped the teacher's interactions and exchanges with the children in each of the sessions. The teacher was fitted with a radio microphone so that her dialogue with children was the focus of data collection. In two of the classes, we videotaped one group of children and audiotaped another group. We also audiotaped each of the teacher meetings.

Given the naturalistic setting and the desire to be as least intrusive as possible, videotaping of children was limited to a focus group in each of two classes. Another focus group in each class was audiotaped. These focus groups were selected after discussions with the teachers and were of mixed achievement levels and gender. One of the main criteria for selecting the focus groups was children's willingness to verbalise while working on tasks.

Other data were collected in response to critical events. That is, the camera would focus on a group who were engaged in resolving some specific aspect of a problem. Other data sources included classroom field notes, children's artefacts (including their written and oral reports), and the children's responses to their peers' feedback in the oral reports.

In our data analysis, we employed ethnomethodological interpretative practices to describe, analyse, and interpret events (Erickson, 1998; Holstein & Gubrium, 1994). This methodological approach allowed us to describe the social world of the classroom by focussing on what the participants said and did, rather than by applying predetermined expectations on the part of researchers. In our analyses, we were especially interested in (a) the nature and development of the mathematical ideas and relationships that the children constructed, represented, and applied; (b) the nature and development of the children's thinking, reasoning, and communication processes; and (c) the development of socio-mathematical interactions taking place within groups (children) and whole-class settings (teacher and children), with particular interest in those interactions involving mathematical argument and justification (Cobb, 2000).

We thus constructed detailed descriptions of the classes to capture the socio-cultural interactions that afford opportunities for children to engage in mathematical learning and reasoning. At a more specific level we used iterative refinement cycles for our videotape analyses of conceptual change in the children (Lesh & Lehrer, 2000). Through repeated and refined analyses of the transcripts and videotapes we were able to identify themes and perspectives that enabled us to make generalisations or assertions about the teachers' and children's behaviours (Cresswell, 1997).

FINDINGS

In reporting our findings, we first address the children's progress on the Butter Beans Problem and then examine their developments on the Airplane Problem. We also consider how children applied their informal, personal knowledge in working the problems.

Butter Beans Problem: Part (a)

Across the four classes, we noted an initial tendency for the children to want to record an answer from the outset, without carefully examining and discussing the problem and its data. The children had to be reminded to think about the given information and share ideas on the problem prior to recording a response. We also observed the children oscillating between analysing the data and discussing at length the conditions required for growing beans. The children drew on their informal knowledge acquired through past experiences in trying to account for the variations in the data. At times, they became bogged down discussing irrelevant issues because their informal knowledge was taking precedence over their task knowledge (i.e., the children's recognition of the specific information presented in the problem). We illustrate this point in later excerpts of the children's work.

We noted at least three approaches that the four classes of children adopted in analysing the data in Table 13.1. The first approach was to focus solely on the results for week 10 and systematically compare rows 1 to 4 for each condition (i.e., compare 13 kg with 15 kg, 14 kg with 14 kg and so on). A variation of this approach was to make the comparisons for each of weeks 6 and 8 as well. A second approach was to add up the data for week 10 in each condition and compare the results. A third but inappropriate variation of the last approach was to sum all of the weights in each table and compare the results. As one child explained, "Sunlight has 146 to 118 (shade). So plants are in sunlight." A further approach (again, inappropriate) was to add the quantities in each row for each condition and compare the end results (i.e., 9kg + 12kg + 13kg for sunlight and 5kg + 9kg + 15kg for shade, and so on).

As the children explored the data initially, they were looking for trends or patterns that would help them make a decision on the more suitable condition. They were puzzled by the anomalies they found and used their informal knowledge to account for this, as can be seen in the following group discussion (hereafter referred to as Amy's group):

Students collectively: 10 against 6, 11 against 10, and 17 against 13.

> **Amy:** So this is obviously better than that, but working out why is the problem.
> **Oscar:** Yes, because the more sunlight the better the beans are. For some reason...
> **Amy:** In some cases, it's less; but in most cases, it's more the same.
> **Tim:** It would depend on what type of dirt it has been planted in.
> **Oscar:** I've got an idea. Perhaps there were more beans in the sunlight.
> **Tim:** We're forgetting one thing. Rain. How much rain!

Amy's group spent quite some time applying their informal knowledge to identify reasons for the trends in data. In doing so, the children engaged in considerable hypothetical reasoning and problem posing, which eventually led them back to a consideration of the task information:

> **Amy:** We're stuck. I can't work this out.
> **Oscar:** I've got an idea. If we didn't have any rain, the sunlight wouldn't ... it wouldn't add up to 17 (kg). And, if we didn't have any sunlight, it wouldn't be up to 17 either. But if we had sunlight and rain....
> **Tim:** Do you want me to jot that down?
> **Oscar:** Don't jot that down because that's wrong. OK, 15 kilograms.

For the remainder of this lesson, Amy's group cycled through applying their informal knowledge to find reasons for why they thought sunlight was better, reviewing the task information by re-examining the sets of data, and attempting to record their findings. In the following excerpt, the group explains to the researchers the dilemma they were facing and the explanations they were considering.

> **Oscar:** But our problem is, we thought it would be because of the rain. It can't get in as well with the shade cloth on. But then we found these results. And we've got a problem. We can't work out why this has popped up. So we're stuck here.
>
> **Amy:** We thought that it was probably that they accidentally put—when they planted the plants, they probably accidentally put slightly bigger plants in this row 1; or the row could have been accidentally longer so it would weigh more. But otherwise, we're sure that sunlight's the best.
>
> **Tim:** I think sunlight's best.
>
> **Researcher:** Why do you think sunlight is better?
>
> **Tim:** Because of the results, like here or here (pointing to week 10 in each condition)
>
> **Amy:** Like, look at 17 to 13 or 18 to 12.
>
> **Researcher:** Or 14 to 14, or 13 to 15.
>
> **Amy:** Yeah, these two are just a bit of a problem, and we've worked out it was probably the row size.

Amy's group made further progress in the next session where they were more focused on the task information with Amy creating a diagram to show the difference in mass between the two conditions (see Figure 13.1). Amy directed her peers' attention to row 3, week 10, where the difference was the greatest ("Here's the best and here's the worst"). Amy attempted to show the other group members this difference by drawing a simple bar graph ("picture graph," as she described it) with the first bar coloured yellow to represent the 18kg (sunlight) and the second bar coloured black to represent the 12kg (shade).

In the excerpt below, Amy is explaining the diagram to the group. At the same time, she is trying to get her peers' attention back onto the problem.

> **Amy:** Ok, guys, if you said this was shade and here's the worst and here's the best (pointing to row 3, week 10)—shade's about there (pointing to her diagram). Here's the best and here's the worst....and that represents the sunlight beans, that would be about the sunlight there (pointing to the yellow shading on her diagram).... what I'm trying to say is the shade is about half as good as sunlight.

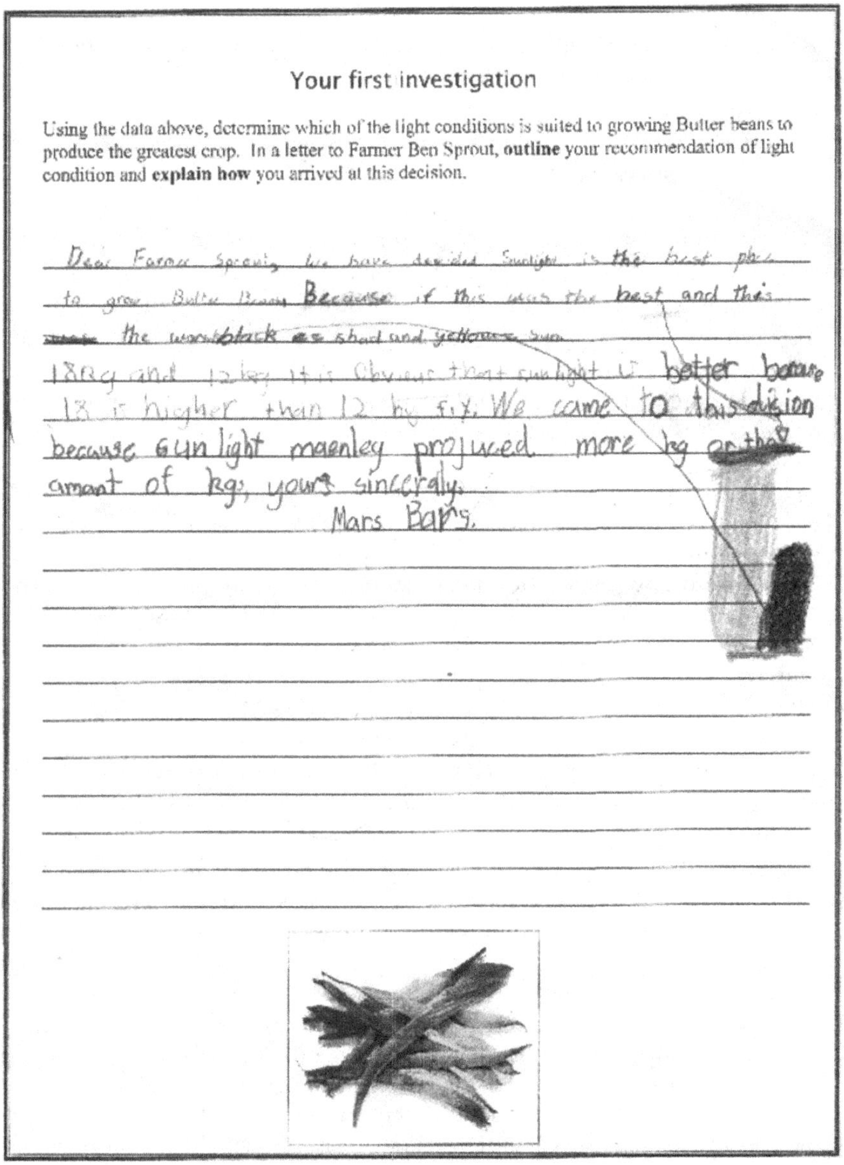

Your first investigation

Using the data above, determine which of the light conditions is suited to growing Butter beans to produce the greatest crop. In a letter to Farmer Ben Sprout, **outline** your recommendation of light condition and **explain how** you arrived at this decision.

Dear Farmer Sprout, we have decided Sunlight is the best place to grow Butter Beans. Because it this was the best and this the world black as shad and yellow sun 18kg and 12kg it is obvious that sunlight is better because 18 is higher than 12 by six. We came to this disision because sun light maenley projuced more kg on the amant of kg. your sincerly.
Mars Bars.

Figure 13.1 Amy's group's representation of the beans' growth in sunlight and shade.

Amy's peers, however, were not listening to her so she decided to pose this question to bring them back on task: "This here is sunlight and this here is shade. Which one's better?" Still not happy with her peers' lack of enthusiasm, Amy posed a more advanced question for her peers:

> **Amy:** Oscar, if this long piece was shade, and the short piece was sunlight, and they represented the weight of the beans, which one would be better?
>
> **Oscar:** This.
>
> **Amy:** No, shade would be because it's bigger. A bigger mass of kilograms.

The difficulty for many of the children across the four classes was completing the letter for Farmer Sprout. As Amy explained to her teacher, "You see, I've drawn a picture graph and we've worked out the answer, but we can't put it into words... I know! We can draw this (her representation) on our letter and explain what it means in words. And that'll get us out of it." The group finally produced the following letter, choosing to focus solely on the largest difference between the conditions:

> Dear Farmer Sprout, We have decided sunlight is the best place to grow Butter Beans. Because if this was the best (an arrow pointing to the representation) and this the worst (another arrow pointing to the representation), black = shade and yellow = sun. 18kg and 12kg. It is obvious that sunlight is better because 18 is higher than 12 by six. We came to this decision because sunlight maenley (sic) projuced (sic) more kg or the amount of kgs. Yours sincerely, Mars Bars (name of the group).

When asked where they obtained their information for this conclusion, Amy explained, "Well, we basically added all of this up (week 10 data for each condition) and we found that shade produced about half as much as sunlight altogether."

Other children produced reports that were embellished with their personal knowledge but limited in reference to task knowledge. For example, a group of boys reported to their class as follows:

> Dear Farmer Sprout. We have measured the conditions that you should grow the butter beans in summer because they will grow better. Butter beans will grow more in the sun than in shade which will make it taste better. They will make you strong. Farmer sprout the beans you are growing are good beans. We think you should pick the beans on Sunday. You should have lots of good beans. Get some spray to kill the bug. Sunlight has 146 kg to 118 kg. So plants, it is in sunlight.

Butter Beans Problem: Part (b)

In responding to the second component of the Butter Beans Problem, the children generally relied on patterns in the data to predict the mass of the beans after 12 weeks. For example, another group in Amy's class reported their predictions for the sunlight condition as follows: "Our findings show that in row 1, week 12, you will get 15 to 17 kilograms, and in week 12, row 2, you'll get 17 kilograms, and in row 3, week 12, you will get 19 to 21 kilograms, and in week 12, row 4, you shall get 18 to 20 kilograms. That's what we think for sunlight." When asked how they got these findings, the children explained, "The data, because we went to week 10 and we counted 2 on . . . because they've sort of gone up like, in twos and it was another two."

When the teacher asked the class if the pattern in each row of the table "was exactly the same, that is, increasing by one or increasing by two," the children agreed that it wasn't. When asked for some reasons why, Amy responded, "Because they're (plants) not made to be a counting pattern." The teacher then discussed with the children various external factors that could be responsible for the different rates of growth.

Children's Responses to the Airplane Problem

As indicated in the Appendix, the Airplane Problem required the children to determine the winner with respect to: (a) The plane that stays in the air for the longest time; (b) The plane that travels the greatest distance in a straight-line path; and (c) The overall winner for the contest. This problem may be considered more challenging than the Butter Beans Problem in that relationships between variables are involved. The Airplane Problem also engages children in a consideration of rules and conditions that anticipate some decision being made.

Across the four classes we observed a variety of approaches to working the problem, with these approaches displaying important mathematical developments. We also noted a few difficulties in the children's interpretation of the table of data and their ways of operating on the data.

On commencing, many children were absorbed in applying their personal knowledge to dealing with the problem. For example, they discussed talking about the nature of the wings, the cabin, the luggage area, and possible flight paths. Some groups physically acted out a plane's flight path, while others made a simple paper plane. We consider this initial discussion and physical representation to be of benefit to the children in familiarising themselves with the problem. Children's application of personal knowledge to this problem was less intrusive than in the Butter Beans Problem, with the exception of the notion of "scratch." Many children associated the term "scratch" with physical marks on a plane, rather than its meaning of elimination. The teacher's intervention was needed here to elicit this alternative meaning from the children.

On continuing with the problem, most groups across the four classes focused on one variable only, be it the number of scratches, the distance travelled, or the time taken by each team. For example, Team C was considered "The winners of the time in the air" and Team E, "The winners of the distance travelled" (the children arrived at these results by adding the respective time and distance data for the three attempts).

There were a few groups who initially operated inappropriately on the data by adding metres to seconds. When probed by their class teacher, one group indicated that they did not fully understand what the data represented, as can be seen in the excerpt below. Notice, however, that Matt had doubted the appropriateness of his group members' actions from the outset.

> **Teacher:** How do you know they would be the winners all the time?
> **Susie:** Because we added up. They are overall winners.
> **Teacher:** Why are they overall winners?
> **Susie:** Because we added up...we added up 32 onto 5.
> **Teacher:** What are these numbers all about? What are you adding up?
> **Matt:** That's what *I* tried to ask them.
> **Teacher:** Well, why don't you look at your labels? The labels are so important.

One group member acknowledged that they had been looking at the labels (units of measure) but responded that "They are the points," indicating that she had difficulty in interpreting the data. Children who added data inappropriately in the Butter Beans Problem (i.e., summing all the weights in each table) also had problems with data interpretation.

Several groups across the four classes initially used the notion of *scratch* as the sole criterion for deciding on possible winners. That is, winners were teams who were not scratched on any trial. The teachers' input here was necessary to challenge this claim. Alex's group, for example, had decided that Team C was the overall winner on the basis that it was not scratched. When the teacher drew attention to the fact that both Team B and Team E had not been scratched, the group quickly reconsidered their answer and stated: "We thought it was either Team C or Team E." On the other hand, children who used the number of scratches as *one* of the criteria for determining the winner revealed elementary probability ideas when they stated that a team had less chance of winning if it were scratched. This understanding is illustrated in the letter of Tom's group, cited later.

We were especially pleased to see children across all four classes develop at least an informal understanding of rate (speed) as they tackled the issue of an overall winner. We provide examples of this development in the following excerpts and begin by returning to Amy's group. In solving the Airplane Problem, Amy explained, "Actually, me and Douglas have worked it out. The people who have the least amount of seconds to the most amount of metres with the least amount of scratches." The teacher asked the group to clarify this statement:

> **Amy:** The least number of seconds with the most metres. So like they spend barely any time flying like 12 metres. They spend one second in 12 metres.
> **Teacher:** So that's one way of looking at it. So you're thinking that it's going to be travelling very fast but a long distance. So would that be to decide the distance travelled?
> **Oscar:** No, the overall.

In later discussion, when Amy's class was presenting their reports, we (the researchers) asked one group of students how their approach to problem solution differed from that of the group who had presented before them. Notice in the discussion below, how a stronger understanding of speed was emerging.

> **Researcher:** An interesting letter. Who can tell me, was the letter that this group wrote ... did it have the same information as the first group's letter, or was it different information?
> **Chris:** Different.
> **Researcher:** In what way was it different?
> **Chris:** Different strategies ... they took notice of the scratches.
> **Researcher:** Anything else different from the first group? (Inaudible student response)
> **Researcher:** Yes, they looked at the least number of seconds, whereas Chris, your group looked at the most number of seconds.
> **Amy:** We thought the least, because it would obviously be a better plane if it could have (inaudible). 13 metres in just 2 seconds means it'd fly really fast rather than say, 13 metres in 20 seconds ... it would be just gliding along. We thought about the speed as well.

In another class, Tom's group explained how they arrived at the overall winner by considering three variables, namely, time, distance, and number of scratches. However, this group considered the greatest time in the air, rather than the least, to be an important variable:

Dear Judges

We have found a way to see who is the winner.

You have to time the team to see who is in the air for the longest.

You have to measure to see who goes the furtherest.

You look closely to see who goes straight and whoever gets the longest gets a
    prize and whoever stays in the air longest gets a prize.

If a team gets scratched, it has less chances.

Whoever gets the longest in the air and the distance is the overall winner.

It could be that the structure of the problem questions influenced Tom's
group (and others) to choose the greatest distance/greatest time relationship when completing their report. We challenged Tom's class by asking
the question, "If two paper planes were thrown and one went 12 metres
in 6 seconds and a second plane went 12 metres in 3 seconds, who would
be the winner?" The children immediately identified the first plane as the
winner, explaining, "Because they (the first plane) stayed in the air longest
and both went the same distance. The six made the difference." There was
agreement with this response across the class.

In yet another class, Menassa's group provided a detailed report that
included the order in which the teams should win and also referred to
an inverse relationship between time in the air and points that should be
awarded. The group members took turns in explaining the system they had
developed:

> **First member:** Longer seconds they take in the air, the less points they
> get. The less time in the air and the longer they go in
> the air, the more points.
>
> **Second member:** Team E was the group you should choose (the child
> made reference to the use of trundle wheels and stop
> watches to measure distance and time respectively). . .
>
> **Third member:** We think that Team E should win the contest. They
> should win because nobody else managed to fly 13
> metres in two seconds. Team A would come second;
> They went 12 metres in 2 seconds. Team B would come
> third; They got 3 seconds in 12 metres, and they had no
> scratches. Team D would come fourth; They got to go
> 12 metres in three seconds but they had one scratch.
> Team C would come fifth because they got 11 metres in
> two seconds and Team F would come last. Team F's best
> score is 11 metres in 2 seconds with one scratch.

One of the researchers queried the group:

Researcher: Would you like to tell us more about those teams? You said
that a team went 12 metres in three seconds. Is that better
than a team that goes 12 metres in six seconds?
Children: Yes, yes.
Researcher: Why did you say that?
Children: Because they took less time in the air.
Researcher: What else were you thinking about?
Children: How far they go.

DISCUSSION AND CONCLUDING POINTS

The modelling problems used in our study encourage young children to
develop important mathematical ideas and processes that they normally
would not meet in the early school curriculum. The mathematical ideas are
embedded within meaningful real-world contexts and are elicited by the
children as they work the problem. Furthermore, children can access these
mathematical ideas at varying levels of sophistication.

In both modelling problems we observed the interplay between children's use of informal, personal knowledge and their knowledge of the
key information in the problem. At times children became absorbed in
applying their personal knowledge to explain the data, which resulted in
slowed progress especially on the Butter Beans Problem. At other times,
children's informal knowledge helped them relate to and identify the important problem information (e.g., understanding the conditions for the
airplane contest). Some groups embellished their written reports with their
informal knowledge, such as referring to additional conditions required for
growing beans. We also observed children recognising when their informal
knowledge was not leading them anywhere and thus reverting their attention to the specific task information. We hypothesise that, in doing so, the
children were showing recognition of and respect for the presentation and
organisation of the data in the problems.

We consider it important that children develop the metacognitive and
critical thinking skills that enable them to distinguish between personal
and task knowledge, and to know when and how to apply each during problem solution. The role of the teacher in developing these skills has been
highlighted by Lehrer and Schauble (2002). That is, teachers need to walk
a tight rope in capitalising on the familiar in data modelling and in "de-

liberately stepping away from it" to assist students in considering the data themselves as objects of reflection (p. 23).

The need to expose young children to mathematical information presented in various formats, including tables of data, is evident from this study. While the children developed facility in interpreting and working with the tables of data, some groups experienced initial difficulties. For example, the cumulative nature of the data in Table 13.1 was not apparent to some children, who added all of the data for sunlight and compared this with the aggregate of the data for shade. Activities in which children collect and record their own data can assist here. In the present study, this was achieved through the preparatory activities leading up to the modelling problems.

In both modelling problems we saw the emergence of important mathematical ideas that the children had not experienced during class instruction. Children's elementary understanding of change and rate of change was evident on both problems, while notions of aggregating and averaging were seen on the Butter Beans Problem. Of particular interest though, is children's informal understanding of speed observed in the Airplane Problem. Some groups focused on the relationship, "shortest time, longest distance" to determine the winning teams, and in so doing, referred to the "speed" of a plane or how "quick/quickly" a plane flew. Other groups considered the relationship, "longest time,longest distance" to be the determinant of the winning plane. The latter could be due in part to the way in which the problem questions were worded. Nevertheless, we see this Airplane Problem as providing opportunities for children to explore quantitative relationships, analyse change, and identify, describe, and compare varying rates of change, as recommended by the NCTM (2000; the Grades 3–5 algebra strand of the *Principles and Standards for School Mathematics*). In addition, we saw elementary probability ideas emerging when children linked the number of scratches with a plane's chances of winning.

Our study has also highlighted the contributions of these modelling activities to young children's development of mathematical description, explanation, justification, and argumentation. Because the problems are inherently social activities, children engage in numerous questions, conjectures, arguments, conflicts, and resolutions as they work towards their final products. Furthermore, when they present their reports to the class they need to respond to questions and critical feedback from their peers. We see this as another area where the teacher's role is important, specifically, in scaffolding the quality of discursive practices.

ACKNOWLEDGEMENTS

This research was supported by a Discovery Grant from the Australian Research Council. The opinions and findings expressed in this article are those of the authors and do not necessarily reflect the view of the Council. We wish to thank our research assistant, Sue Mahoney, for her assistance in developing the activities of our study, and the third-grade teachers and their classes for participating in our study. Their contributions are greatly appreciated.

Reprint of article that appeared in a Special Issue of the *Mathematics Education Research Journal* (Early Childhood Education), November 2004.

REFERENCES

Batterham, R.. (2000). *The chance to change.* The final report by the Chief Scientist. Canberra: Department of Industry Science and Resource.

Carpenter, T. P., & Romberg, T. A. (2004). *Powerful practices in mathematics & science: Research-based practices for teaching and learning.* Madison: University of Wisconsin.

Cobb, P. (2000). Conducting teacher experiments in collaboration with teachers. In R. A. Lesh & A. Kelly (Eds.), *Handbook of research design in mathematics and science education* (pp. 307–334). Mahwah, NJ: Lawrence Erlbaum.

Cresswell, J. W. (1997). *Qualitative inquiry and research design: Choosing among five traditions.* Thousand Oaks, CA: SAGE.

Diezmann, C. M., & Watters, J. J. (2003). The importance of challenging tasks for mathematically gifted students. *Gifted and Talented International, 17*(2), 76–84.

Diezmann, C., Watters, J.J., & English, L. D. (2002). Teacher behaviours that influence young children's reasoning. In A. Cockburn & E. Nardi (Eds.), *Proceedings of the 26th International PME Conference* (pp. 289–296). Norwich: University of East Anglia.

Dockett, S., & Perry, B. (2001). "Air is a kind of wind": Argumentation and the construction of knowledge. In S. Reifel & M. Brown (Eds.), *Early Education and Care, and Reconceptualizing Play, 11,* 227–256.

Doerr, H. M., & English, L. D. (2003). A modeling perspective on students' mathematical reasoning about data. *Journal for Research in Mathematics Education, 34*(2), 110–136.

English, L. D. (2002). Priority themes and issues in international mathematics education research. In English, L. D. (Ed.), *Handbook of international research in mathematics education* (pp. 3–16). Mahwah, New Jersey: Lawrence Erlbaum.

English, L. D. (2003). Reconciling theory, research, and practice: A models and modelling perspective. *Educational Studies in Mathematics, 54,* 2 & 3, 225–248.

Erickson, F. (1998). Qualitative research methods for science education. In B. J. Fraser & K. G. Tobin (Eds.), *International handbook of science education* (Part 2). (pp. 1155–1173). Dordrecht: Kluwer Academic Publishing.

Gainsburg, J. (2004, submitted). *The mathematical modelling of structural engineers.* Paper submitted for publication.

Holstein, J. A., & Gubrium, J. F. (1994). Phenomenology, ethnomethodology and interpretitive practice. In N. K. Denzin & Y. S. Lincoln (Eds.), *Handbook of qualitative research* (pp. 262–285). Thousand Oaks, CA: SAGE.

Inhelder, B., & Piaget, J. (1958). The growth of logical thinking in from childhood to adolescence. In H. E. Gruber & J. J. Voneceh (Eds.), *The essential Piaget.* New York: Basic Books.

Jones, G., Langrall, C., Thornton, C., & Nisbet, S. (2002). Elementary school children's access to powerful mathematical ideas. In L. D. English (Ed), *Handbook of International Research in Mathematics Education* (pp. 113–142). Mahwah, NJ: Lawrence Erlbaum.

Lamon, S. (2003). Beyond constructivism: An improved fitness metaphor for the acquisition of mathematical knowledge. In R. A. Lesh, & H. M. Doerr (2003). *Beyond constructivism: Models and modelling perspectives on mathematics problem solving, learning, and teaching* (pp. 435–448). Mahwah, NJ: Lawrence Erlbaum.

Lehrer, R., Giles, N. D., & Schauble, L. (2002). Children's work with data. In R. Lehrer & L. Schauble (Eds.), *Investigating real data in the classroom: Expanding children's understanding of math and science* (pp. 1–26). New York: Teachers College Press.

Lehrer, R., & Schauble, L. (2003). Origins and Evolutions of Model-based Reasoning in Mathematics and Science. In R. A. Lesh & H. Doerr (Eds.), *Beyond constructivism: Models and modeling perspectives on mathematics problem solving, learning, and teaching* (pp. 59–70). Mahwah, NJ: Lawrence Erlbaum.

Lesh, R. (1998). The NEXT STEP Project. Purdue University. Unpublished research document.

Lesh, R., & Doerr, H. M. (2003). (Eds.). *Beyond constructivism: Models and modeling perspectives on mathematics problem solving, learning, and teaching.* Mahwah, NJ: Lawrence Erlbaum.

Lesh, R., & Harel, G. (2003). Problem solving, modelling, and local conceptual development. *Mathematical Thinking and Learning, 5,* 157–190.

Lesh, R. & Heger, M. (2001). Mathematical abilities that are most needed for success beyond school in a technology based age of information. *The New Zealand Mathematics Magazine, 38*(2), 1–17.

Lesh, R. & Lehrer, R. (2000). Iterative refinement cycles for videotape analyses of conceptual change. In R. Lesh & A. Kelly (Eds.), *Research design in mathematics and science education.* Hillsdale, NJ: Lawrence Erlbaum.

National Council of Teachers of Mathematics (2000). *Principles and standards for school mathematics.* Reston, VA: National Council of Teachers of Mathematics.

O'Connor J., White J., Greenwood, P., & Mousley, J. (2001). Enabling sciences still on the slippery slide. Press Release on behalf of the AIP, the RACI, the AMS, the AMSC and the IEA [http://www.aip.org.au/initiative2002/index.html Accessed 2/9/2004].

Perry, B., & Dockett, S. (2002). Young children's access to powerful mathematical ideas. In L. D. English (Ed.), *Handbook of international research in mathematics education.* Mahwah, NJ: Lawrence Erlbaum.

Richardson, K. (2004). *A design of useful implementation principles for the development, diffusion, and appropriation of knowledge in mathematics classrooms.* Unpublished doctoral dissertation, Purdue University.

Steen, L. A. (Ed.). (2001). *Mathematics and democracy: The case for quantitative literacy.* National Council on Education and the Disciplines. USA.

Stillman, G. (1998). The emperor's new clothes? Teaching and assessment of mathematical applications at the senior secondary level. In P. L. Galbraith, W. Blum, G. Booker, & I. Huntley (Eds.), *Mathematical modelling: Teaching and assessment in a technology-rich world* (pp. 243–254). West Sussex: Horwood Publishing Ltd.

Tate, W., & Rousseau, C. (2002). Access and opportunity: The political and social context of mathematics education. In L. D. English (Ed), *Handbook of International Research in Mathematics Education* (pp. 271–300). Mahwah, NJ: Lawrence Erlbaum.

Vygotsky, L.S. (1978). *Mind in Society.* Cambridge, MA: Harvard University Press.

Yackel, E., & Cobb, P. (1996). Sociomathematical norms, argumentation, and autonomy in mathematics. *Journal for Research in Mathematics Education, 27,* 458–477.

Zawojewski, J. S, Lesh, R., & English, L. D. (2003). A models and modelling perspective on the role of small group learning. In R. A. Lesh & H. Doerr (Eds.), *Beyond constructivism: Models and modeling perspectives on mathematics problem solving, learning, and teaching* (pp. 337–358). Mahwah, NJ: Lawrence Erlbaum.

APPENDIX

THE INDOOROOPILLY TIMES

Students fly away in the Annual Paper Airplane Contest at local school

If the Wright Brothers, pilots, and aircraft engineers can do it, surely the students in Indooroopilly State School's year three classes can do it.

What will you be doing, that a couple of inventors, some of the best pilots in the world, and the brightest minds in the world do everyday? Fly!

You will attempt to be like the Wright Brothers and design an airplane that will meet today's airplane standards.

However, you won't be using aluminium, various metal parts or jet engines for these planes. All you will need are pieces of paper – or any other craft materials – and a whole lot of imagination.

You have the opportunity to design planes that will be able to fly long distances. In the contest, you will need to design a plane that will travel in a straight path.

However, with every contest there is a set of rules that you must follow to try to win the contest's grand prize. Some of these rules are:
1. No cuts can be made in the plane's wings,
2. Parts may be cut off from the plane entirely, and
3. You must build your own planes.

You will be working in groups to design and test your planes before contest day. Each group gets *three attempts*.

Scratches may occur in this contest. A scratch means that the plane did not travel in a straight path for any of the flight.

I have heard that some of you are getting way into this – someone said that you are bringing in the in-flight refreshments! This will be an interesting contest.

Reflection Questions:

1. What is the Annual Paper Airplane Contest about?

   _____

   _____

   _____

   _____

2. What needs to be done to design an airplane that will be successful for the contest?

   _____

   _____

   _____

   _____

   _____

   _____

3. What does it mean if your plane is scratched in one of your attempts?

   _____

   _____

   _____

4. What units of measurements are used in contests in which distance and time are measured?

   _____

   _____

The Annual Paper Airplane Contest

This year, the Stoney Creek State School will hold their annual paper airplane flying contest on 15th September. Students in year three will be working in groups and will design one plane.

All planes will be designed to fly for as long as possible in the air (*time*) and over a long *distance* from a target. The plane will need to travel in a *straight-line path*.

Three awards will be given at this contest. One will be given to the group whose plane stays in the air the longest – another to the group whose plane travels the longest straight-line path – and the final award is an overall award given to the group who wins the contest.

Results from the Annual Paper Airplane Contest 2002

Team	Attempts	Time in the Air (seconds)	Distance travelled in a straight path (metres)
Team A	1	2	11
	2	2	12
	3	scratch	scratch
Team B	1	3	12
	2	1	7
	3	1	8
Team C	1	1	9
	2	3	11
	3	2	11
Team D	1	3	12
	2	scratch	scratch
	3	1	8
Team E	1	2	9
	2	1	10
	3	2	13
Team F	1	1	9
	2	2	11
	3	scratch	scratch

Investigation

In the past years, the judges have had problems with deciding how to select a winner and how to judge the contest. Using the given data from the previous years, find a way to help the judges decide on the overall winner of the contest.

Write a letter to the judges of the contest explaining to them how to determine who wins each of the categories (time in the air and distance travelled in a straight-line path) and how to decide the winner of the overall award for the contest.

SECTION V

TECHNOLOGY AND THE NET GENERATION

CHAPTER 14

CONNECTED GIFTEDNESS

Mathematical Problem Solving by Means of a Web Technology: Case of the CASMI Project

Viktor Freiman and Nicole Lirette-Pitre
Université de Moncton

INTRODUCTION

In the wave of exponential development of the virtual Web, the world becomes more and more connected through the extensive use of new technology tools. Social networks are getting more complex and rich by social links that were unthinkable even a decade ago. Very often, such social networks take the form of new communities built of people having common goals and interests that produce and share collaboratively new information (knowledge) that becomes almost instantly available to all members and sometimes to the whole worldwide Internet community. This phenomenon of virtual social networking is becoming the object of reflection, analysis, and research; several authors are trying to make sense of what happens in this new world and how it affects our society and each individual. A remark made by Rennie and Mason (2004, p. 5) looks at the Web as a 'mixing bowl

Interdisciplinarity, Creativity, and Learning, pages 205–216
Copyright © 2009 by Information Age Publishing
205

and takes the mixture into virtual format that has more tenuous relationship with time and space'. Continuing their metaphor, authors argue that 'ubiquity, speed and global scale provide the heat to cook this into something we have not tasted before' (Rennie, & Mason, 2004, p. 5) .

Educators have a particular interest in this topic because of its everyday impact on the life and learning of young generations that goes beyond a simple use of technology but rather lives with technology being 'connected'. Being connected goes far beyond a simple (even extensive) use of web technologies to search for new information or to communicate with others. It is also knowing how to instantly get (a click away) 'the right information at the right time from anybody and anywhere' (Veen, & Vrakking, 2006, p. 30). The same authors analyze abilities of so-called *Homo Sapiens* to learn at a very early stage that there are many different sources of information that may claim different truths, so they need to learn how to filter the information and make up their mind in networks of peers to whom they communicate frequently (Veen, & Vrakking, 2006). How does this affect our understanding of giftedness? In our chapter, we will discuss this question in more details developing on the paper written by Blanchard (see the chapter in this volume). We introduce the term "connected" giftedness in order to describe new pedagogical options that appear in the way we teach and learn. As an example of a new virtual environment that may foster connected giftedness, we analyze a virtual community that offers opportunities of online mathematical problem solving to the thousands of students and hundreds of university students enrolled in mathematics education courses.

Several authors point at the importance of providing mathematically gifted students with more challenging and complex problems (Cline, 1999; Sheffield, 2003; Mann, 2005; Freiman, 2006). Research also shows that there is a lack of pedagogical resources to engage these students in cognitively more complex activities, hence making them construct new knowledge (Silver, 1998). Students in our classrooms are still solving a series of routine application problems that can be resolved by only applying a ready-to-make mathematical concept, operation or formula, and do not require a higher order of thinking and thus, are less cognitively and meta-cognitively demanding (Diezmann & Watters, 2004). What can be done to bring more challenging, open ended, investigative tasks that are needed to support genuine learning within gifted and all other learners?

Assouline & Lupkowski-Shoplik (2003) list several options for developing mathematical talent in- and outside the classroom. It includes breadths and depth approach, enrichment topics, math-related independent study projects, curriculum compacting, telescoping, subject-mater acceleration and ability grouping. The later list contains mathematical competitions and clubs, summer programs, weekend programs, individually placed programs, magnet schools, and distance learning and correspondence courses.

In our chapter, we will look at yet another option of mathematics enriched learning supported by technology. This would bring a new dimension in the learning "beyond the classroom." We argue that online learning represents a new spectrum of experiences for mathematically gifted students connecting mathematics to their natural (virtual) environment and eventually helping them develope their creativity (see the chapter written by Manuel in this volume) in a virtual collective solution space that can be generated by online problem solving activities.

An explosion of e-resources in the last 5 years makes possible developing of multiple online forms of knowledge building and sharing in new informal dynamic and interactive ways that are impossible in traditional in- or extracurricular settings. In our chapter, we are going to analyze what online learning can give to the education of mathematically gifted students. We are going to use an example of a virtual mathematical and scientific problem solving community that we created called CASMI (Communauté d'Apprentissages Scientifiques et Mathématiques Interactifs, www.umoncton.ca/casmi).

Analyzing connections between mathematical giftedness and creativity, Freiman and Sriraman (2007) argue that the nurturing environment suitable for both mathematical giftedness and creativity should allow students to choose and use representations that would enhance ability to model a problem and to resolve problems by developing an attitude of building a network of new questions, new resolutions, going out of initial problems, and to communicate using different tools of communication. Such environment an would combine, according to Meissner (2005) individual and social components (motivation, curiosity, self-confidence, flexibility, engagement, humor, etc.), challenge (fascination, excitement, provoking, even thrilling), and important abilities (to explore, structure, invent, modify, listen, argue, cooperate, etc.). In many ways, our analysis is built on those ideas.

We will start bye presenting a brief theoretical overview of virtual enriched learning opportunities and environments that support and enhance it.

VIRTUAL LEARNING COMMUNITIES IN MATHEMATICS

According to Jonassen et al. (2008), technologies are not teachers or repositories of information but rich and flexible media for representing what students know and what they are learning. Davitt (2005) stresses the role of information communication technology (ICT) in accelerated learning process as it lowers threat and can provide second and third chance learning. This also provides different points of access for learners with different learning styles, allowing teachers to create, store and deploy multimedia resources and activities to work with different visual, auditory and kinaesthetic modalities, provide new ways of making connections, and provide support and scaffolding. All this may help all students to learn better, especially the gifted ones. Klotz (2003), at his turn, affirms that in mathematics, as in other disciplines, the Web is expanding our concept of the classroom itself, changing what is learned and how it is learned. The term of virtual schooling is used to introduce a notion of the "anytime/anyplace" learning, to develop new opportunities for independent learning and dramatically reconfigure the work of both teachers and learners (Hunter, & Smith, 2001).

Enrichment and differentiation are mentioned most often in the literature that analyses mathematical problem solving in virtual environments. Piggot (2004) argues that on-line resources are not suitable for only the most able ones but offer new learning opportunities for pupils of nearly all abilities, and that enrichment is not only an issue of content but also a particular teaching approach that offers opportunities for exploration, discovery and communication. Renninger & Shumar (2002), at their turn, put much emphasis on the idea of interactions between members of a virtual community that interact around the services and resources generated by all participants thus providing a basis for participant knowledge building about mathematics, pedagogy, and/or technology and developing a culture that encourages collaborative problem posing and problem solving. Upon Pallascio (2003), virtual environments like *Agora de Pythagore* (http://euler. cyberscol.qc.ca/pythagore/) can contribute to the creation of a mathematical research community where students discuss several philosophical questions online. Questions like *what is infinity?*, in which they will be actors and also creators of their own knowledge via argumentative discourse.

WHAT IS CASMI?

The primary goal of the CASMI site is to develop and to improve problem solving culture in mathematics and science in K–12 students. In defining this culture, we put emphasis on two particular abilities: to reason and to communicate mathematically at the highest competence levels[1] (PISA, 2003). Both of these correspond to the main didactical principles of the recent New Brunswick K–12 mathematics curriculum's framework that brings mathematical development above and beyond the specific learning outcomes by proposing mathematically complex and contextually rich problems called "situations-problèmes" (MENB, 2003). This is supposed to maximize the students' intellectual potential while taking into account individual needs and pace of learning including those students who have above average abilities.

Bi-weekly, twelve problems appear on the CASMI website, four problems for each subject: mathematics, science and chess. Participants can choose

Figure 14.1 The CASMI web page where students can choose which problems he wants to solve, selecting one of the 4 animals (*Manchot* [Penguin], *Girafe* [Giraffe] *Dauphin* [Dolphin], and *Hibou* [Owl]) under the discipline in which he wants to solve (mathematics, science or chess).

and solve as many problems as they like. Mathematical problems posted on the web site present a variety of mathematical contexts that can often be interpreted in more than one way and require a deeper understanding of mathematical concepts to solve them.

These problems are based on real life related situations and context or may present purely mathematical contexts. To solve them, the students must be able to translate their interpretation into mathematical models which require more than one mathematical concept and more complex cognitive and meta-cognitive strategies. This is what brings intellectual challenges that gifted students need.

All problems, once they are posted, become available through the archive any time, anywhere to every member of the virtual community. It becomes an important issue for teachers and students to have permanent free access to high-speed Internet which would allow everyone to work on CASMI at their own pace; save their incomplete solutions in their portfolio and come back anytime they wish until the problem is inactivated. Having one to one access to the Internet would be very beneficial for gifted students who are known to have a better understanding and quicker grasp of mathematical concepts and usually finish their work faster. Using the CASMI problems online, they and can find a problem upon their level and their personal taste and thus get an opportunity for extra challenges that are unavailable in the regular classroom.

Our aim is not to select gifted students but rather offer a variety of problems, so that every student *including the gifted ones* to connect to a socially friendly, mathematically rich and technologically supported network so contributing to the construction of collective knowledge and to the maximization of her own intellectual potential.

COMMUNICATION INITIATED BY CASMI

The concept of the community aims to link people together inviting everybody to become a registered member of one of three groups in our community: students, university students and teachers. One of the benefits of becoming a member is free access to a personal virtual portfolio (e-portfolio) as well as to the common cyberspace in which everyone can choose and solve a problem, look for its analysis and the most interesting solutions that were submitted. They can also create problems and consult any problem from the archive (using a search engine if necessary) and browse through interesting external educational links. All these communication activities help to support a network for interactive and collaborative online learning (Rennie & Mason, 2004).

By using a dynamic database structure, we created a possibility for each schoolchild who sends us a solution (by using an electronic form integrated in CASMI) to get formative feed-back from university students taking a mathematical education class. By means of this feedback, the students get a better understanding of their successes and also recommendations of what they would need to improve. Knowing that the solution will be analyzed by someone else than the classroom teachers makes students feel more comfortable and brings additional motivation (Freiman, Manuel, 2007).

According to our research data (Freiman, Manuel, 207), this particular virtual interaction between students and university students is appreciated by the students and teachers and reflects a technology provoked phenomenon known as 'breaking the walls of the classroom, bringing learning beyond school curriculum" (Freiman, Lirette-Pitre & Manuel, 2007).

Moreover, the participating school teachers can get access to their student's portfolios and follow their progress further. They can organize diverse classroom enrichment activities with the whole class or with a group of students using the database of problems available on the site. CASMI also offers a discussion forum where participants can ask questions, share their opinions and have discussions about the problems that are posted as well as other subjects of interest. Every member can propose a subject of discussion and post a comment on any other topic.

Even without direct contact between members of the virtual community an intensive virtual communication creates interpersonal links and becomes an important part of the online leaning process.

MENTORING PROBLEM SOLVING PROCESS: UNDERSTANDING AND FACILITATING STUDENTS' MATHEMATICAL DEVELOPMENT

As we have mentioned above, an important aspect that makes the CASMI website interactive is that each online solution submitted is evaluated formatively by university students enrolled in mathematics education classes. When analyzing students solutions, we aim to give them a positive and informal feedback that will make them feel encouraged even if their solution is incorrect. At the same time, it is important that this feed-back provides students with the detailed formative analysis of their strategies and communication skills.

Using a special electronic form, we assess how students retrieve, interpret, and represent the necessary data, the strategy they use to solve the problem and how they implement and explain it allowing another person to understand their solution. Another important component to look at is the meta-cognitive reflection on the whole problem solving process which allows the student to validate the results, to think of other possible answers or other possible strategies to solve the problem, and finally to pose a new problem. Some of our evaluation criteria can be chosen from several predefined options, others can be added as a free comment. This approach brings an advantage of providing every student with a written feed-back using pictorial icons (star, sun, book, etc.) and personally addressed comments. At the same time, some students may prefer a verbally given feedback, or even more instant interaction with the "virtual mentor."

By communicating virtually with real students and reflecting on their experience in classroom discussions and course work, our future teachers learn how to understand children's reasoning and communication skills and learn how to provide them with an appropriate assessment. Each university student is invited to solve the problem themselves, which makes them experience the same process as the students. At the same time, they should anticipate different ways to solve the problem and children's difficulties and possible obstacles to overcome.

One of the most important outcomes of this experience is that pre-service teachers discover new ways of teaching and learning that is more open to the variety of styles and interpretations along with various possible uses of online resources to enhance their teaching. The need for such openness and flexibility will be demonstrated by the concrete example of one CASMI problem.

CONNECTED GIFTEDNESS: PATTERNS OF PARTICIPATION AT THE CASMI PROJECT

Putting problems online and let students solve them is only the first step of connecting giftedness and technology. It is a day after day, month after month year after year work of implementation, monitoring and development of social activities that actually make giftedness connected. That is why participation patterns are important indictors of how the new community realizes its common goals.

The CASMI virtual community was activated in October 2006. Table 14.1 shows the number of participants in May 2007 (after one school year). Remarkably at the end of the first year, the community grew to over 5000 participants within a short period. Most of the participants are from New-Brunswick, Canada which can be easily understood by the success of its previous CAMI version, in-school presentations and workshops for teachers done by the CASMI team in local districts. But we can also see some interest from the Quebec teachers and students. There were however few attempts to promote the project outside of the New Brunswick.

The second table (Table 14.2) presents the number of solutions submitted by subject (math, science, chess) and by category (manchot, girafe, dauphin, and hibou). It also shows a number of correct solutions. The overwhelming number of solutions on mathematics problems can by explained by the culture developed and supported by teachers during 6 years of the CAMI project (which had only mathematical problems), many students explore the science and chess sections on their own. There is no surprise that in mathematics, the number of solutions vary more from the "easiest" level of difficulty (manchot) to the "highest" (hibou). In science and chess, the variation is not that big. If we look at the proportion of correct solutions, in all three subjects, it decreases whereas the difficulty level increases.

TABLE 14.1 CASMI Participants' Statistics

Country (province)	Teachers	University students	Students
Canada:	616	220	4475
NB	491	212	4048
Quebec	114	7	361
Other provinces	11	1	66
International:	113		1
France	99		
Other countries (9)	14		
Total:	5,425 participants		

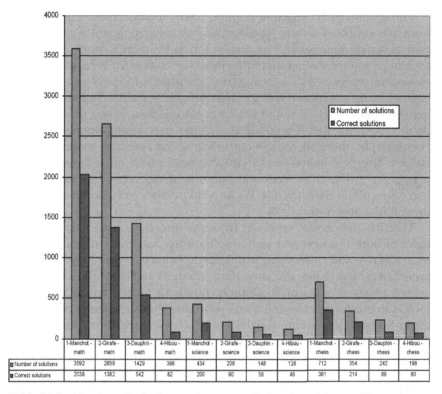

	1-Manchot - math	2-Girafe - math	3-Dauphin - math	4-Hibou - math	1-Manchot - science	2-Girafe - science	3-Dauphin - science	4-Hibou - science	1-Manchot - chess	2-Girafe - chess	3-Dauphin - chess	4-Hibou - chess
Number of solutions	3592	2658	1429	386	434	208	148	126	712	354	242	198
Correct solutions	2038	1382	542	82	200	90	58	46	361	214	89	80

Table 14.2   Number of solutions and correct solutions by subject and by category.

While all participants must have a name and password, other information such as school and grade level can not be verified. Moreover, during the first year, the registration form contained the school level as a "string of symbols," so it could not be treated automatically. Following the qualitative observation of the level mentioned in each submitted solutions, we can only suppose that most students are from the middle school Grades (6–7–8). In the second year, we modified our registration form in which the school level is now the number, we also added the age as a mandatory field, so we will be able to get these data soon.

We were also able to learn from our conversation with workshop participants (teachers and administrators) that there are different kinds of participation, such as teacher required use, open sessions during class time and participation at home. Another pattern yet to be studied is that of students who "hop" from one problem level and kind to another. They spend little time on any one problem even if they are getting them right or wrong. Other students are more persistent and will spend more time on a problem but do only one problem (often assigned to them as a school work).

The study of such behaviour requires construction of more sophisticated tools and probably on-site (in-classroom) investigation which will be our goal in the coming years. A discussion forum provides a room for student exchanges to topics suggested by members or by the CASMI team. The discussion forum opened the community discussions to ideas and questions beyond mathematics and science. However, some partial data analyzed using the online discussion and from several solutions to the problems students solved revealed several interesting patterns among the participants. For example, there seems to be a strong social aspect to some students' entries. They add happy faces to their answers and want to know who the people are giving them feedback. They ask what kind of movies one likes and tell us about their friends. We hope to pursue our in-depth enquiry of these initial observations.

The first theme that emerged from the interviews is an overall excitement of both teachers and students about the Internet environment as a source of rich and challenging problems provided by technology. While this same fact is neither new nor surprising (teachers and students were already actively engaged in the project), their comments suggest several reasons to look at in our further investigation : good variety of problems and contexts going beyond the minimum fixed in the school curriculum, possibility of choice of the problem according to student's learning needs and personal interest, nice and attractive (and interactive) environment comparatively to traditional types of resources, and thee possibility of getting a formative feed-back from pre-service teachers (not from YOUR teacher). In addition to this, some socio-affective factors emerge as well: intrinsic motivation, proud of success, risk-free environment with no assessment related sanctions and stress, possibility of sharing discoveries with peers and teachers. It is also to mention that 1 to 1 access to the laptops and wireless Internet in the classroom facilitated the harmonious use of the resources which seems to be another important factor contributed to positive response.

Another issue that comes from our findings is related to the problem of differentiation in the mathematics classroom. As we said in our previous sections, today's mathematics curriculum recognizes differences in how children learn and the right of each child to receive an education that is adapted to their needs. The New Brunswick math curriculum stresses that in order to meet educational needs of each student; teachers have to use a variety of approaches. The virtual environment presents one possible resource that gives each pupil a chance to choose an appropriate problem, solve it at their own pace using their own strategy and communication tool. It also brings some informal elements in the classroom routine. The fact that each participant gets personal feedback from the university student can be seen as a motivating factor for students because they see that their work brings attention to other people and is being socially valorized by personal attention or even public recognition (children can see this recognition when their solution is posted as interesting or their name is placed on the congratulation list). However, more research is needed in order to study in more details the factors of intrinsic motivation among participants of the project.

REFERENCES

Assouline, S., & Lupkowski-Shoplik, A. (2003). *Developing Mathematical Talent: A Guide for Teachers and Parents of Gifte Students,* Purfrock Press, Inc., 387 pp.

Cline (1999). *Giftedness has many faces: multiple talents and abilities in the classroom.* The Foundation of Concepts in Education, Inc., 193 pp.

Davitt, J. (2005). *New Tools for Learning.* Network Educational Press. 141 pp.

Diezmann, C., Watters, J. (2004). *Challenge and connectedness in the mathematics classroom: using lateral strategies with gifted elementary students.* In: Proceedings of the Topics Study Group 4: Activities and Programs for Gifted Students. The 10th International Congress on Mathematical Education, July 4–11, 2004, Copenhagen, Denmark/ Eds. E. Barbeau, H. Shin and others, University of Latvia, Riga.

Freiman, V., Manuel, D et Lirette-Pitre, N. (2007). CASMI Virtual Learning Collaborative Environment for Mathematical Enrichment. Understanding our Gifted, summer 2007, pp. 20–23.

Freiman, V., & Manuel, D. (2007). Apprentissage des mathématiques. Dans S. Blain, et al. *Les effets de l'utilisation de l'ordinateur portatif individuel sur l'apprentissage et l'enseignement.* Rapport final. Présenté au MENB par le CRDE et l'équipe de recherche ADOP, Université de Moncton, Mars, 2007.

Freiman, V., & Sriraman, B. (2007). *Does mathematics education need a working philosophy of creativity?* In: Mediterranean Journal for Research in Mathematics Education, Vol. 6, No. 1&2, 2007, pp. 23–46.

Jonassen, D., Howland, J., Marra, R., & Crismond, D. (2008). Meaningful Learning with Teachnology (3rd edition). Pearson, 253 pp.

Hunter, & Smith (2001). *Virtual Schooling: Integrating Schooling into Technology.* In: B. Barrell (Ed.). Technology, Teaching, and Learning: Issues in the Integration of Technology, Detselig, pp. 197–220.

Klotz, G. (2003). Math: Calculating the Benefits of Cybersessions. In: D. T. Gordon (Ed.) The Digital Classroom: how technology is changing the way we teach and learn. Harvard Education Letter.

Mann, E. (2005). Dissertation: Mathematical Creativity and School Mathematics: Indicators of Mathematical Creativity in Middle School Students. University of Connecticut, 2005, www.gifted.uconn.edu/siegle/Dissertations/Eric%20 Mann.pdf

Meissner, H. (2005) *Creativity and mathematics education.* Paper presented at The 3rd East Asia Regional Conference on Mathematics Education http://www.math. ecnu.edu.cn/earcome3/sym1/sym104.pdf

MENB (2003). Programme d'études en mathématiques. Nouveau-Brunswick.

Pallascio, R. (2003). *L'agora de Pythagore : une communauté virtuelle philosophique sur les mathématiques*. Dans: Taurisson, A., Éd. Pédagogie.net: l'essor de communautés virtuelles d'apprentissage. PUQ.

Piggott, J. (2004). Mathematics Enrichment: what is it and who is it for? Retrieved 11.02.2009. at http://nrich.maths.org/public/viewer.php?obj_id=2719

OCDE (2000). *Measuring Student Knowledge and Skills*: A New Framework for Assessment; http://www.pisa.oecd.org/dataoecd/45/32/33693997.pdf

Rennie, F. & Mason, R. (2004) *The Connection: Learning for the connected generation.* IAP, 170 pp.

Renninger, K. Ann, and Wesley Shumar (Eds) (2002). *Building Virtual Communities: Learning and Change in Cyberspace.* Cambridge, England: Cambridge University Press.

Sheffield, L. (2003). *Extending the challenge in mathematics.* TAGT & Corwin Press, 150 pp.

Veen, W., Vrakking, B. (2006). *Homo Zappiens: Growing up in a digital age.* Network Continuum. 160 pp.

NOTE

1. Here is how PISA defines the highest level of competency:

   ....students can conceptualise, generalise, and utilise information based on their investigations and modelling of complex problem situations. They can link different information sources and representations and flexibly translate among them. Students at this level are capable of advanced mathematical thinking and reasoning. These students can apply this insight and understanding along with a mastery of symbolic formal mathematical operations and relationships to develop new approaches and strategies for attacking novel situations. Students at this level can formulate and precisely communicate their actions and reflections regarding their findings, interpretations, arguments, and the appropriateness of these to the original situations.

   Retrieved November 2, 2009, from http://www.scotland.gov.uk/Publications/2004/12/20390/48514

CHAPTER 15

TEACHING AND LEARNING FOR THE NET GENERATION
A Robotic-Based Learning Approach

Samuel Blanchard
University of Moncton

INTRODUCTION

Let's begin by taking a break from everyday life to rekindle childhood memories. . . . What were your favourite activities? What were your past-times? What kind of games did you and your friends play? How was school growing up? What was the classroom environment? How did you learn? What did you learn? What kind of school activities did your teachers organize in order to make you learn? What were the tools suggested to you by your teachers to learn? More likely, those activities and tools helped you become the person you are today. It taught you social and cognitive skills that you are, in all probability, using in your everyday life.

Now take a moment to think about today's generation of students! It can be your own children, your nieces or nephews, cousins, whoever it may be! Can we say that this generation of children is similar to yours? How so? Do they live the same way? Do they, in general, have the same childhood activ-

Interdisciplinarity, Creativity, and Learning, pages 217–231

217

ities? Do they play with the same toys? Do they have the same skills as you did at their age? Is their school the same? Do you think the teachers teach the same way? Do you think they learn in the same manner?

It's evident that today's society is in a state of evolution. Since the informational and technological industrial revolution (Greenwood, 1999), our civilization has never been the same. Technologies have become a great part of our everyday life and most likely will for generations to come. Tapscott clearly states in his latest book *Grown Up Digital—How the Net Generation is Changing Your World* that "If you look back over the last 20 years, clearly the most significant change affecting youth is the rise of the computer, the Internet, and other digital technologies" (Bennett, Maton, & Kervin, 2008; Prensky, 2001a; Tapscott, 2009). However, how does technology shape our youth generation in learning?

This article will try to answer these questions using a multitude of literature on technology and the corresponding links to learning. This article will, in a general manner, bring you into the world of today's generation of students: The Net Generation (Tapscott, 2009) or Digital Natives (Prensky, 2001a) and its connections to an innovative pedagogical technique using technology as its highest cognitive potential with Robotic-based learning.

It is important to specify that there is evidence of a lack of empirical research regarding today's learners (Bennett et al., 2008) and the "Net Generation" or "Digital Natives." They have a need for an evolution of pedagogical framework based on visionary concepts founded by observatory assumptions (supported by weak quantitative and qualitative data analysis) by numerous researchers (Bennett et al., 2008). However, there is clear evidence from our daily lives that the new generation of students were born and grew up in a different world than their predecessors resulting in different experiences. Moreover, Ormrod (2008), clearly states that learning as a "long-term change in mental representations or associations as a result of experience" making way to behavioural changes and brain development (as a result of brain plasticity) (Ormrod, 2008). Is it safe to assume that a different brain development through experience (in this case, technology-based evolution) has a non-negligible cause–effect relationship on pedagogical needs for optimal learning? Not necessarily so, but the concept of Net Generation or Digital Natives is a possible and, in my opinion, a plausible hypothesis to counter the consequences of such cause–effect on the pedagogical framework regarding the needs of the new generation of students. One thing is certain, further research is clearly needed.

NET GENERATION: DEFINITION[1]

Today's young children are the first to be more knowledgeable, comfortable and literate then their parents (Tapscott, 2009). What made them different from their close relatives? The Net Generation (Tapscott, 2009) or the digital Natives (Prensky, 2001a) are a first of its kind and are defined by their upbringing filled with the use of new technologies from an early age (Barnes, Marateo, & Ferris, 2008; Marbito & Medley, 2008; Oblinger & Oblinger, 2005; Prensky, 2001a). They represent 27% of the total population of the United States, where born between 1977 and 1997 and are the baby-boomers (born 1946–1964) offsprings (also known as the baby-boomers "echo" generation), or the digital Immigrants (Oblinger & Oblinger, 2005; Prensky, 2001a; Tapscott, 2009).

Because of the digital extensions and enhancements that we perceived today, Prensky, in 2009, states that as the years go by, the differences between the Natives and the Immigrants will be less noticeable paving way to a less dichotomous nomenclature to a more modern interdisciplinary grey scale called digital wisdom. Digital wisdom is a combination of a person's conceptive ability in a technological as well as general wisdom. As one develops the other will enhance accordingly (Prensky, 2009). As technology advances, the current students develop digital wisdom at the same rate as the technological advancement suggesting that the digital wisdom is self perpetuated.

In this technology filled world, by the time students reach college, each will have played over 10 000 hours of video games, sent and received over 200 000 e-mails and instant messages, spent over 10 000 hours talking on the digital cell phones, spent over 20 000 hours spent watching TV and seen over 500 000 commercials (Prensky, 2001b). A non-negligible experience that may change children's learning needs and perceptions and a definite non-negligible difference in their parent's use of technology which was known as the TV generation (Oblinger & Oblinger, 2005; Tapscott, 2009).

Everything about these kids' lives is different. Firstly, the family dynamics is a lot different from the previous families; the authority has shifted so that each member's values and opinions is valorised (Tapscott, 2009) Secondly, the Net Generation seems to be smarter than the previous generation with a three point IQ average increase per decade since World War II (Tapscott, 2009; Viadero, 2002). Thirdly, the Net Generation is a generation that wants to learn and value education and its purpose for a successful future (Barnes et al., 2008). Lastly, the Net Generation have different sets of generational values and tend to socialize with different people (age, gender, social class, race/ethnic background, sexual orientation, etc.) with tolerance (Tapscott, 2009):

> While there are important differences across cultures, nations, genders and classes, the evidence is strong that this is a positive generation, with strong values. Take young people in North America—they care about the world. They are open, tolerant, and least prejudiced generation ever. (Tapscott, 2009)

What does this all mean? What impact does this generation have on schools today? Are our schools ready? Does this generation change the role of teachers compared to previous generations? As teachers, how can we assure an optimal learning experience for these students?

NET GENERATION: THE EIGHT NORMS AND PEDAGOGICAL PRINCIPALS

The role of a teacher is to create healthy and creative environments where students can reach their unique potential as citizens of the world regardless of physical or mental deficiencies, gender, age, race/ethnic background, or any other minorities that might have a psychological or physical impact on the student's ability to learn. Fundamentally, it's the teacher's responsibility and obligation to create several pedagogical scenarios that takes into consideration the needs of all students and ensures a flawless delivery with the help of the school environment (school direction, all students, teachers assistant and school personnel) and the community to help all pupils to become a unique citizen in the world. In conjunction with the definition of the Net Generation, it is clear that teachers have a tough job to perform with today's students which have a series of characteristics and needs that unifies them into a new group of citizens. This new group is different from that of the past generations of students which, in return, demands a different pedagogical paradigm than what the traditional setting can offer.

Many authors tend to suggest that it is apparent that schools today, as its normal function, may not be designed to teach the Net Generation (Barnes et al., 2008; Oblinger & Hagner, 2005; Oblinger & Oblinger, 2005; Prensky, 2001a, 2001b, 2005; Tapscott, 2009). Therefore, we can ask ourselves, what makes the Net Generation so different in a learning perspective from the other generations and what does it entail as the pedagogical principles?. The following section is dedicated to explain the eight norms of the Net Generation (Tapscott, 2009), the pedagogical evolution needed to meet the standards brought on by the eight norms (Sword & Leggott, 2007; Tapscott, 2009) and the new pedagogical principles (Sword & Leggott, 2007; Tapscott, 2009) a teacher might want to follow to ensure the Net Genera-

Teaching and Learning for the Net Generation

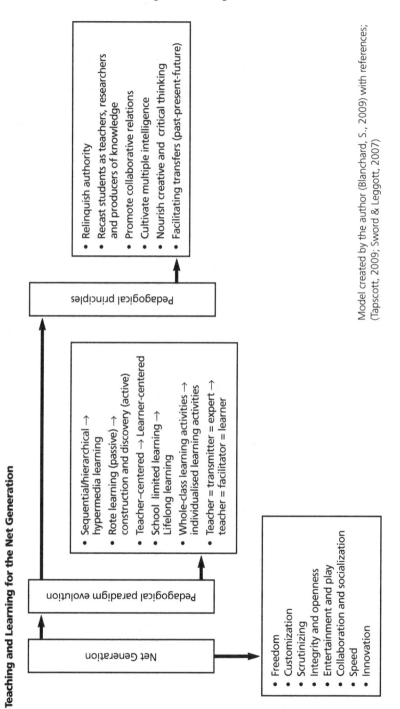

Net Generation

- Freedom
- Customization
- Scrutinizing
- Integrity and openness
- Entertainment and play
- Collaboration and socialization
- Speed
- Innovation

Pedagogical paradigm evolution

- Sequential/hierarchical → hypermedia learning
- Rote learning (passive) → construction and discovery (active)
- Teacher–centered → Learner-centered
- School limited learning → Lifelong learning
- Whole-class learning activities → individualised learning activities
- Teacher = transmitter = expert → teacher = facilitator = learner

Pedagogical principles

- Relinquish authority
- Recast students as teachers, researchers and producers of knowledge
- Promote collaborative relations
- Cultivate multiple intelligence
- Nourish creative and critical thinking
- Facilitating transfers (past-present-future)

Model created by the author (Blanchard, S., 2009) with references;
(Tapscott, 2009; Sword & Leggott, 2007)

tion has an inclusive environment in the classroom to optimise the learning experience of all.

First of all, the changes needed to teach the Net Generation in our current school systems stems from the eight norms of the Net Generation; which is part of a whole new paradigm of evolution, a new way of living brought on by technologies. Educating the Net Generation is not an easy task when you are a digital immigrant or underdeveloped Digital Wisdom. The eight norms are exactly that, the adaptation that the digital immigrants tend to make and the abilities the new generation, the Net Generation or the Digital Natives, have developed (part of Digital Wisdom) and assumed since youngsters.

Freedom is what I want and need!

One aspect which is very important to the Net Generation is choice. They are asking teachers to give them the liberty to live and experience. Making mistakes is part of life, and they are able to know that mistakes are the pathway to learning. Freedom of expression is also a must and thinking and living outside the box of conformity is more and more apparent. "Young people insist on freedom of choice. It's a basic feature of their media-diet. Instead of listening to the top 10 hits on the radio, the Net Geners compose iPod playlist of thousands of songs chosen from millions of tunes available" (Tapscott, 2009). It is also true that "the internet has given them the freedom to choose what to buy, where to work, when to do things like buy a book or talk to friends, and even who they want to be" (Tapscott, 2009). What does this mean in the classroom?

The Net Generation, with the help of broadband Internet, have access to the Web on demand. This feature gives them the opportunity to have access to millions of informational websites being able to answer their questions when and where they want it (Tapscott, 2009). What does this mean to regular classrooms with passive lectures, the ideal pedagogical framework from a traditional classroom (Tapscott, 2009)?

Customization is the way of expressing myself!

In this day and age, with the help of the Web 2.0, every internet user is able to customize their personal working space to reflect who they are and showing it to whomever is interested. "Millions around the world don't just access the Web, they are creating it by creating online content" (Tapscott, 2009), via Web 2.0 websites such as Wikipedia (www.wikipedia.org), Facebook (www.facebook.com), Flickr (www.flickr.com), blogs and wikis. It is a portal to discovering ourselves in this complex world and having our own piece of the puzzle on the web is valorising ones' work by making it accessible to the world to see, a true motivational success!

Scrutinizing is seeing, by our better judgement, what is not meant to be seen!

Doeswhat I see, really exist or possible? That is often a question one might ask themselves if they are from the Net Generation. To them, it is important to question the information presented to them. "Given the large number of information sources on the Web, not to mention unreliable information – spam, phishers, inaccuracies, hoaxes, scams, and misinterpretations–today's youth have the ability to distinguish between fact and fiction" (Tapscott, 2009).

Corporate integrity and openness? It's what I'm looking for to make my career decisions!

In a sense, integrity and openness is a way to prevent mistrust brought upon scrutiny. "Integrity, to the Net Geners, primarily means telling the truth and living up to your communications" (Tapscott, 2009). "They want their universities, schools, government, and politicians to be honest, considerate of their interests, accountable, and open" (Tapscott, 2009). In other words, it's important for a person to keep their ideas transparent so mistrust can be prevented. What does this mean for teachers and school personnel? A high level of transparency is a must! It is important to be upfront and honest to students by letting them know what your intentions are and the need for a specific pedagogical scenario. Being honest with oneself and the classroom is, by far, a key to success with the Net Generation.

Entertainment and play: Keep me motivated and I'll learn, trust me!

"This generation has been bred on interactive experiences" (Tapscott, 2009). Today's youth like and need to play and have fun doing it. Video games has been responsible for giving them the opportunity to play and learn at the same time while developing problem-solving skills and being a great source of motivation (Gee, 2003; Royle, 2008).

Net Geners are also known for their multitasking ability (Oser, 2005). They often do their homework while listening to music, chatting with their friends on instant messaging programs like MSN and browsing through the Internet.

It is also believed that the Net Generation has a small attention span in result of a high order of stimuli provided by technologies paving way to motivational issues in the classroom (Barnes et al., 2008). However, we can all agree that by watching kids play video games, watching movies, listening to music or simply surfing the internet we rarely observe such attention span (Prensky, 2001b, 2005)! What is so different with their school environment? Why can't school work be as entertaining?

Collaboration and socialization: Bring forces together for great power!

"The Net Generation are natural collaborators" (Tapscott, 2009). They work together by communicating in synchronous (eg. MSN, Skype) and asynchronous (eg. Forums, blogs) communication devices or programs, in multi-payer online gaming to resolve a certain task or by sharing files via online communities (eg.Desire2Learn, wiki) in an effort to combine strong abilities or compensating for less then perfect ones to create a highly potential community (Oblinger & Oblinger, 2005; Tapscott, 2009).

> The current model of pedagogy is teacher focused, one-way, one size fits all. It isolates the student in the learning process. Many Net Geners learn more from collaborating-both with their teacher and with each other. They'll respond to the new model of education that's beginning to surface-student-focused and multiway, which is customized and collaborative. (Tapscott, 2009)

Speed: I want it NOW!

Growing up with technological-evolution, the time needed to find information has diminished tremendously. The Net Generation demand answers right away and this also implies in their school work (Barnes et al., 2008). They appreciate continual performance retroactions (Barnes et al., 2008; Oblinger & Oblinger, 2005) to ensure a rapid learning continuum and self-esteem (Barnes et al., 2008; Tapscott, 2009).

Innovation: we are going somewhere new!

Net Geners grew up in a world of constant innovation. They want their workplace and school to be as innovative and creative (Tapscott, 2009).

"Net Geners told us that an innovative work environment is perceived to be leading edge, dynamic, creative and efficient" (Tapscott, 2009). Thus, they want and expect high end technologies in the classroom so they are able to work and learn as they are meant to.

To conclude, the Net Generation differs from their parents by wanting and needing freedom to express themselves, to customize their surrounding in a form of expression, by scrutinizing all information given to them in an attempt to verify information, by wanting the people around them to have integrity and openness to their values and goals, by needing to collaborate and socialize, by demanding speed and constant gratification in all area of their lives and trying to innovate and live in a environment that innovates with time. Again, what does this mean to teachers and the educational system?

THE EVOLUTION: FROM THE NET GENERATION TO NEW PEDAGOGICAL PRINCIPLES

The Net Generation is growing up in a tool-rich environment and this needs to be taken into account in designing pedagogical systems
—Philip, 2007

In an attempt to give answers to the possible cause–effect resulting from the presence of the Net Generation in our classrooms on the current pedagogical paradigms, some authors, namely Tapscott (2009) and Sword and Leggott (2007), have tried to elaborate some key points to work on as a community of teachers and school personnel.

Before elaborating on some Net Generation friendly pedagogical principles, it is important to understand a series of pedagogical paradigm evolutions brought on by the characteristics and needs of today's students. Firstly, it is important for our classroom to change from a sequential or hierarchical learning method (provided by a more traditional approach) to a more hypermedia learning (Prensky, 2001b). Today's learner tend to have "hypertext mind" (Prensky, 2001b) developed by technologies, meaning they tend to make more leaps and multiple links between specific concepts and to an extent, different subjects. Secondly, teachers need to keep students active in their learning environment (Oblinger & Oblinger, 2005). Unlike traditional learning methods, which are focused on memorization and drilling, also known as rote learning (Wikipedia, 2008), the Net Geners' brain prefers pedagogical scenarios based on a more open-based approach and that provides them multiple opportunities to do some hands-on learning (Oblinger & Oblinger, 2005). In addition, an active and individualised learning paradigm makes way to a more learner-centered approach where students have more control over their learning outcomes (in consequence of: their work paste, their interest, their specific objectives in career choices, etc.) and are the main focus when designing learning activities. It is also important for teachers and schools to promote lifelong learning with learning outcomes dedicated to life skills and the introduction of informal learning opportunities with technologies like the CASMI environment (www.umoncton.ca/casmi). Lastly, these pedagogical principles demand the teachers to have a different role in the classroom. They become facilitators making way to co-construction with the entire learning community which consist of the school personnel, the community and most importantly, the students.

This series of pedagogical paradigm evolution cannot stand on its own without a series of pedagogical principles that needs to be put into place to promote optimal learning. Using the pedagogical evolution, the section bellow will briefly give you indications on how it is possible to build a rich environment for the Net Generation's students.

The first thing a teacher needs to do is relinquish authority (Mabrito & Medley, 2008; Sword & Leggott, 2007)! It is of first importance because without this relinquishment, it will be almost impossible to change ones philosophical concept of education in today's world. In order to fully integrate technology in the classroom, the only factor against change is the teacher's attitude and pedagogical paradigm (Depover, Karsenti, & Komis, 2008). Things need to change in the classroom to promote a Net Geners friendly learning environment.

> We know much more than our students do. But they know much more then we do. When we renounce our own exclusivity status as erudite experts, placing students in the role of teachers and ourselves in the role of students, not only do we model for them the benefits of life-long learning, but we allow them to experience firsthand what every seasoned teacher already knows: If you really want to master a subject, teach it. (Sword & Leggott, 2007)

In a second attempt to make today's children learn, it is important to recast them as teachers, researchers and producers of knowledge (Sword & Leggott, 2007) with the help of an open-based pedagogy like Problembased learning (PBL).

> Problem-based learning can be defined as an inquiry process that resolves questions, curiosities, doubts, and uncertainties about complex phenomena in life. A problem is any doubt, difficulty, or uncertainty that invites or needs some kind of resolution. (Barell, 2007)

PBL, is based on challenging students to become highly motivated and involved as knowledge researchers, teachers and producers in an attempt to answer to their own questions (Barell, 2007; Goodnough & Cashion, 2006;

Kanter & Schreck, 2006; B. C. Nelson & Ketelhut, 2007; T. H. Nelson, 2004). This authenticity described by Barrel (2007), is a catalysis of transferable abilities and knowledge; a life-long learning experience (Ormrod, 2008).

It is also important to promote collaborative relations with an excellent cooperative pedagogy (Sword & Leggott, 2007) "The ability to work within different groups, as well as mutual understanding, are proving to be increasingly important as we face the challenges of modern life arising from, among other things, the growth of societal diversity" (Gamble, 2002). In general, corporation in education is a form of learning organization which permits small heterogeneous groups of students to meet the learning outcomes while being interdependent while fully participating with members of the learning community (Gaudet, 1998).

In addition to letting go of authority, making a learner-centered classroom and promoting collaboration, it's important to cultivate multiple intelligence, or individualized learning. Pedagogical differentiation is learner-centered individualized approach of learning (Caron, 2003; Tomlinson, 2004). More specifically, Tomlinson (2004), dictates that a differentiated classroom is a place where teachers make sure each student is in constant competition with themselves, all the while taking into account the learning history of each student (Tomlinson, 2004). Essentially, teachers who practice a differentiated classroom suppose that each student, from gifted to learning disabilities, have common learning abilities but they also have non-negligible differences which characterize them as learners. Therefore, teachers practicing in this pedagogy adopt and assume such philosophy and plan in consequence (Tomlinson, 2004).

Learning outcomes should always be contextualized paving way to knowledge transfers. It is important for students to make links between the past, their present to prepare themselves to the future (Sword & Leggott, 2007). An easy way to make links between knowledge is to practice interdisciplinary in the classroom to show students how someone can make links between general and specific contents within all disciplines.

Last but not lease the importance of creativity and critical thinking in the classroom. Tomorrows workers will be asked to resolve problems and conflicts and use technologies that the teachers of today cannot imagine (Sword & Leggott, 2007). Consequently, it's important to give them the opportunity to think, ask questions, debate certain issues and promote the use of divergent thinking in problem solving and every day activity. When it comes to technology use, certain applications are known to be more cognitively powerful; demanding a more creative and critical thinking. Depover (2008) stipulates that there is three kinds of technologies that demand different complexity of cognitive thinking while using it; Teaching assisted by technologies (passive learning), learning assisted by technologies (active learning) and experimentation assisted by technologies (hands-on active learning) (Depover et al., 2008). It is clear that going from "teaching" to "learning" to "experimenting," students have a different role as learners and researcher. Since the Net Geners tend to appreciate a more innovative learning space, wants to be motivated at all time and likes hand-on learning, the experimentation assisted by technologies would be an excellent learning activity.

To conclude, a question one might ask would be how can one teacher integrate these principles in the classroom? The next section will give you an idea of the type of activities possible with the use of robotics in the classroom.

TECHNOLOGICAL INNOVATION: ROBOTIC-BASED LEARNING (RBL)

The use of robotics in the classroom is, by its nature, an exciting technological breakthrough and relatively unused in classrooms today (Williams, Ma, Prejean, & Ford, 2007). Papert, in 1980, started his robotics research using a constructionism approach to give students a chance to interact with new technologies (Papert, 1980, 1991). "Constructionism is a learning and instructional theory rooted in the philosophical tradition of constructivism, which emphasizes the active role of the learner in collaboratively constructing knowledge in a rich context" (Duffy & Cunningham, 1996).

Since this pedagogical and technological innovation, a series of scientist have attempted to answer a series of research questions in robotic-based learning. Let's begin by saying that robotics projects had a positive effect on science and technology motivation in classrooms (Barker & Ansorge, 2007; Carbonaro, Rex, & Chambers, 2004; Gura, 2007; Nourbakhsh et al., 2005; Petre & Price, 2004; Williams et al., 2007). In addition, more research have shown a positive effect on the level of collaboration between students (Nourbakhsh et al., 2005; Petre & Price, 2004), on problem solving and critical thinking skills in children (Mauch, 2001; Norton, McRobbie, & Ginns, 2007; Petre & Price, 2004; Wagner, 1998), on motivation and interest level of students (Mauch, 2001; Nourbakhsh et al., 2005; Petre & Price, 2004), on the ability to use inquiry skills in classroom (Williams et al., 2007), on the learning of a programming language (Nourbakhsh et al., 2005), while being an excellent interdisciplinary agent (Papert, 1980; Rogers & Portsmore, 2004). Even if such papers have indicated a strong pertinence in such an approach in the classroom, there is still a need for further research in robotic-based learning to ensure its use in K–12 setting all around the world (Barker & Ansorge, 2007; Williams et al., 2007).

Furthermore, Robotic-based learning does concord directly with most, if not all, pedagogical principles listed in the previous section. Firstly, Robotic-based learning is a high potential tool utilizing Inquiry because of its discovery-based pedagogical philosophy resulting in acquisition of problem-solving skills (Williams et al., 2007). A lot of researchers dictate that problem-solving skills, which comes from a unique mind-set in each child resulting from experience, can be obtained with the help of a discovery-based learning technique like problem-based learning (Barell, 2007; Ormrod, 2008) which is in direct parallel to the pedagogical evolution and three pedagogical principles (relinquish authority, recast students as teachers, researchers and producers of knowledge, promote multiple intelligences and nourish creative and critical thinking) recommended for the Net Generation. Although the use of Robotic-based learning is effective in promoting multiple intelligences in students, there is yet to be significant research on the use of this technique within the gifted and learning disabled students. Secondly, Robotic-based learning is an interdisciplinary catalyst resulting in transferable knowledge which helps the students to understand the value of such activity for their cognitive and social development and future careers. Robotics-based learning, lastly, promotes collaborative relations between the students and their teacher in the classroom resulting in a powerful team for learning: we work together, we learn together!

CONCLUSION

Alas, it is safe to assume that this Robotic-based learning is a Net Generation friendly approach and should be utilized in its full potential from an early age in every school that values the identity and high potential learning in students and wants to prepare a generation of citizens ready to solve today's and tomorrow's local, nationwide and international social and scientific issues.

BIBLIOGRAPHY

Barell, J. (2007). *Problem-based learning: an inquiry approach* (2nd ed.). Thousand Oaks: Corwin Press.

Barker, B. S., & Ansorge, J. (2007). Robotics as means to increase achievement scores in an informal learning environment. *Journal of Research on Technology in Education, 39*(3), 229–243.

Barnes, K., Marateo, R., & Ferris, P. (2008). Teaching and Learning with the Net Geneation. *Educause Review, 3*(4).

Bennett, S., Maton, K., & Kervin, L. (2008). The "digital natives" debate: A critical review of the evidence. *British Journal of Educational Psychology, 39*(5), 774–786.

Carbonaro, M., Rex, M., & Chambers, J. (2004). Using LEGO Robotics in a Projects-Based Learning Envirnment. *Interactive Multimedia Electronic Joural of Computer-emhanced learning.*

Caron, J. (2003). *Apprivoiser les différences.* Montréal: Les Éditions de la Chenelière.

Depover, C., Karsenti, T., & Komis, V. (2008). *Enseigner avec les technologies–favoriser les apprentissafes, développer des compétences.* Québec, QC: Presses de l'Université du Quévec.

Duffy, T. M., & Cunningham, D. J. (1996). Construictivism: Implications for the design and delivery instruction. In D. H. Jonassen (Ed.), *Handbook of research for Educational Communications and Technology* (pp. 170–198). Bloomington, IN: Association of Educational Communications and Technology.

Gamble, J. (2002). Pour une pédagogie de la coopération. *Éducation et francophonie, 30*(2), 186–217.

Gaudet, D. (1998). *La coopération en classe.* Montreal, QC: Chemelière/McGraw-Hill.

Gee, J. (2003). *What video games have to teach us about learning and literacy.* London: Plagrave Macmillan.

Goodnough, K., & Cashion, M. (2006). Exploring Problem-Based Learning in the Context of High School Science: Design and Implementation Issues. *School Science and Mathematics, 106*(7), 280.

Greenwood, J. (1999). The Third Industrial Revolution: Technology, Productivity and Income Inequality. *Economic Review (Federal Reserve Bank of Cleveland), 35*(2).

Gura, M. (2007). Student Robotic. In *Classroom Robotics: Case Stories of 21st Century Instruction for Millennial Students* (pp. 11–31). Charlotte: Information Age Publishing.

Kanter, D. E., & Schreck, M. (2006). Learning Content Using Complex Data in Project-Based Science: An Example from High School Biology in Urban Classrooms. *New Directions for Teaching and Learning, 2006*(108), 77–91.

Mabrito, M., & Medley, R. (2008). Professor Johnny can't read: Understanding the Net Generation's texts. *Innovate, 4*(6).

Marbito, M., & Medley, R. (2008). Why professor Johnny Can't Read: Understanding the Net Generation's Texts. *Educause Revire, 4*(6).

Mauch, E. (2001). Using technological innovation to improve the problem-solving skills of middle school students: Educator's experiences with the LEGO Mindstorms Robotic Invention System. *The Clearing House, 74*(4), 211–213.

Nelson, B. C., & Ketelhut, D. J. (2007). Scientific Inquiry in Educational Multi-User Virtual Environments. *Educational Psychology Review, 19*(3), 265–283.

Nelson, T. H. (2004). Helping Students Make Connections. *Science Teacher, 71*(3), 32–35.

Norton, S. J., McRobbie, C. J., & Ginns, I. S. (2007). Problem solving in a middle school robotics design classroom. *Research in Science Education, 37*, 261–277.

Nourbakhsh, I. R., Crowley, K., Bhave, A., Hsium, T., Hammer, E., & Perez-Bergquist, A. (2005). The robotic autonomy mobile robotics course: Robot design, curriculum design and educational assesment. *Autonomous Robots, 18*(1), 103–127.

Oblinger, D., & Hagner, P. (2005). *Seminar on educating the Net Generation.* Paper presented at the EDUCAUSE, Tempe, AZ.

Oblinger, D., & Oblinger, J. (2005). Is it Age or IT: First Steps Toward Understanding the Net Generation. In *Educating the Net Generation*: Educause.

Ormrod, J. E. (2008). *Human Learning* (5 ed.). Upper Saddle River, New Jersey: Pearson Education.

Oser, K. (2005). Kids cram more hours in media day. *Advertising Age, 76*(46).

Papert, S. (1980). *Mindstorms: children, computers, and powerful ideas.* New-York, NY: Basic Books.

Papert, S. (1991). Situating constructionism. In I. Hartel & S. Papert (Eds.), Constructionism Available from http://www.papert.org/articles/SituatingConstructionism.html

Petre, M., & Price, B. (2004). Using robotics to motivate "back door" learning. *Education and Information Technologies, 9*(2), 147–158.

Philip, D. (2007). The Knowledge building paradigm: A model of learning for Net Generation students. *Innovate, 3*(5).

Prensky, M. (2001a). Digital Natives, Digital Immigrants. *On the Horizon, 9*(5).

Prensky, M. (2001b). Digital Natives, Digital Immigrants: De they really Think Differently? *On the Horizon, 9*(6).

Prensky, M. (2005). "Engage me or Enrage Me"—What today's learners demand. *Educause 40*(5), 61–64.

Prensky, M. (2009). H. Sapiens Digital: From Digital Immigrants and Digital Natives to Digital Wisdom. *Innovate, 5*(3).

Rogers, C., & Portsmore, M. (2004). Bringing engineering to elementary school. *Journal of STEM Education, 5*(3 et 4), 17–28.

Royle, K. (2008). Game-Based Learning: A different perspective. *Innovate, 4*(4).

Sword, H., & Leggott, M. (2007). Backward into the future: Seven principles for education the Ne(x)t generation,. *Innovate, 3*(5).

Tapscott, D. (2009). *Grown up digital–how the Net Generation is changing your world.* New York City: McGraw-Hill.

Tomlinson, C. (2004). *La classe différenciée.* Montréal: Chenelière Education.

Viadero, A. (2002). Nature vs. nurture = starting jump in IQs'. *Education Week, 21*(19).

Wagner, S. P. (1998). Robotics and children: Science achievement and problem solving. *Journal of Computing in Childhood Education, 9*(2), 149–192.

Wikipedia. (2008). Rote Learning. Retrieved November 17th, 2008

Williams, D. C., Ma, Y., Prejean, L., & Ford, M. J. (2007). Acquisition of physics content knowledge and scientific inquiry skills in a robotics summer camp. *Journal of Research on Technology in Education, 40*(2), 201–216.

NOTE

1. Most information from the Net Generation are based upon Tapscott (2009) book called *Grown Up Digital—How the Net Generation is Changing Your World.*

CHAPTER 16

DOES TECHNOLOGY HELP BUILDING MORE CREATIVE MATHEMATICAL ENVIRONMENTS?

Dominic Manuel
University of Moncton

INTRODUCTION

In today's world, the societies deal with many problems locally, nationally and even internationally, issues that require more complex and interdisciplinary thinking. How does this affect our school curricula and teaching and learning practices that aim to prepare future generations of students? How is the complexity of those issues related to the knowledge and skills that are taught in and beyond the classroom settings? What should be learned in school in order to be able to solve today's most open and complex problems? The concept of creativity has been brought into the school system in many contexts related to these questions, even in disciplines that are traditionally considered as less open to the variety of strategies and approaches as it happens in mathematics.

In her synthesis of discussions held during the last ICME Congress, Sheffield (2008) resumes that in the twentieth century, as technology progresses

Interdisciplinarity, Creativity, and Learning, pages 233–247
Copyright © 2009 by Information Age Publishing
All rights of reproduction in any form reserved.

rapidly, even though computation and memorizing rote procedures are important abilities, it is more important to help students become more creative problem solvers by nurturing and cultivating the abilities to recognize and define problems, generate multiple solutions or paths toward solutions, reason, justify conclusions, and communicate results. Also, the 2000 NCTM *Principles and Standards for School Mathematics*, emphasize a combination of abilities that allow including students in more creative mathematical work through building new mathematical knowledge through problem solving, making and investigating mathematical conjectures, understanding how mathematical ideas interconnect and building on one another to produce a coherent whole, and creating and using representations to organize, record and communicate mathematical ideas (NCTM, 2000).

According to Chan (2008), teachers tend to present mathematical contents to the best of their knowledge and control how students learn it in a unique way without letting students create their own sense of the content. Other authors analyze how teachers believe that for each problem, there is only one solution and one way of solving it, as well as their willing to teach explicitly "good" strategies in order to solve a specific problem may lead to students' responding by applying one simple algorithm and in many cases, the last one learned in class in order to please their teacher by finding the single correct answer that she expects (Hashimoto, 1997, Poirier, 2001). Sometimes emphasis is also put on speed and accuracy as characteristics of a gifted student (Mann, 2005).

In summary, the pedagogy used inside the classroom seems to emphasize what students reproduce instead of how students think. By doing so, students may loose their natural curiosity and become less enthusiastic towards mathematics, which can also make it difficult for teachers that do want to develop mathematical creativity in their classroom (Meissner, 2005). How to help students, especially those who are gifted develop mathematical creativity?

This chapter will focus on answering the previous question based on a literature review. Also, results of an exploratory research will be presented on mathematical creativity using the CASMI virtual community. The choice of a virtual community for this particular study was based upon the fact that there appears to lack researches in such an environment. Different definitions and models of creativity will first be acknowledged based on a literature review.

MATHEMATICAL CREATIVITY: RESEARCH PERSPECTIVES

According to Sriraman (2004), the first known time in history that the term mathematical creativity appeared was in 1902 in a francophone periodical named *L'enseignement des mathématiques* (Teaching mathematics) by a mathematician named Henri Poincaré. Researches on this concept were not popular from that time up to the first half of the century. In more recent times researchers are getting more interested in the topic.

For a long time, creativity was often cited along with artistic creations, musical compositions and scientific discovery. Thus, mathematics wasn't known to be a topic associated with the concept (Chan, 2008). In fact, this term was mostly used in relation to the education of gifted students. Renzulli's (1998) 3-rings model of giftedness (Figure 16.1) considers giftedness as the intersection between above average abilities, task commitment and creativity where creativity is composed of fluency (finding different possible answers), flexibility (finding different possible strategies to a problem), insight and originality of ideas and strategies in problem solving and also the ability to create new problems (Renzulli, 1998).

Other conceptions of creativity focus on a production process which requires long periods of reflection and experimentations that any student can develop with considerable time and effort while solving non-familiar problems, a process that can be long, flexible and deep (Holyoak & Thagard, 1997; Stemberg & Lubart, 1996). This vision gives creativity different dimensions than being just a subset of giftedness according to Renzulli's model (Meissner, 2008; Treffinger & Isaksen, 2005).

What are characteristics of mathematical creativity? How creativity can be detected and/or developed in mathematics? The next paragraphs will discuss existing theoretical views on these questions.

There are numerous ways to define mathematical creativity. Over 100 definitions can be found in literature (Mann, 2005). For example, Runco (1993) described creativity as a multifaceted construct that involves con-

Figure 16.1 3-ring model of giftedness of Renzulli (1998).

vergent and divergent thinking, problem posing and problem solving, self-expression, intrinsic motivation, a questioning attitude and self-confidence. Haylock (1987) talked about the ability to find new relationships between techniques and areas of application and to make associations between possibly unrelated ideas. Krutetskii (1976) used the contexts of problem formation, invention, independence and originality to characterize mathematical creativity. Others have applied the concept of fluency, flexibility and originality to the concept of mathematical creativity (D. Haylock, 1997; Mann, 2005). Singh (1988) defined mathematical creativity as the process of formulation of hypotheses concerning cause and effect in a mathematical situation, verifying them multiple times to make modifications and conclusions and then communicating the results. Also, some authors focus on two main aspects: the originality and the utility of ideas (Anabile, 1996; Feist, 1998; Stemberg & Lubart, 1999).

Social and cultural dimensions of creativity can be found in so-called "confluence" approaches such as "systems approach" (Csikszentmihalyi, 2000), "the case study as evolving systems approach" (Gruber & Wallace, 2000), and the "investment theory approach (Stemberg & Lubart, 1996)

Gruber and Wallace (2000) also argue in their theory that creative work is always the result of purposeful behavior. However, Sriraman (2004) argues that the discovery of penicillin as a counterexample to that statement. Can we consider accidently made illuminations as creative work? The resulting product might be classified as creative or innovative, but can we characterize the behavioral process as creativity? How can we also relate creativity to the problem posing and solving which remain commonly accepted forms of productive (and potentially creative) mathematical activity?

In the following section, we will present different strategies that can be implemented in order to develop mathematical creativity focusing on problem solving.

DEVELOPING MATHEMATICAL CREATIVITY

Summarizing literature review, we claim that mathematical creativity occurs when students get the opportunity to find different possible and original solutions and strategies to a given problem, take risks, and sometimes try to find new relationships between facts or ideas.

In order to help promote mathematical creativity, Cline (1999), Freiman (2006), Freiman & Sriraman (2007) propose using rich mathematical tasks in the form of open-ended problems especially when it comes to the education of gifted and talented students. Takahashi (2000) defines open-ended problems as problems that can be solved using multiple strategies and can have different possible solutions depending on the individual's interpretation of the problem. Two of the biggest advantages of these problems are the fact that they tend to disconnect us from the stereotype that every problem has only one solution and that students can solve those problems upon their abilities (Klavir & Hershkovitz, 2008). Open-ended problems are therefore more challenging cognitively because they open the doors to multiple interpretations, strategies and possible solutions which brings students the opportunities to construct their knowledge in a wide variety of life contexts. These types of problems allow the student to confront the same problem in different perspectives and present all the mathematical concepts and relations in different ways until they discover a strategy that will let them further investigate the problem while they value their own inventions and make their own discoveries. Solving these types of problems let the students reach the first steps towards mathematical creativity (Mann, 2006).

Looking at problem solving as collective work, Leikin (2007) uses the term *collective solution space* a group of solutions that is built by a group of problem solvers that produces collectively different solutions to a given problem. According to the same author, collective solution spaces enable the development of individual solution spaces (solutions done by one student) by means of interpersonal communications. When a collective solution space is gathered, it is possible to expose them in order to create discussions where all the participants can share their mathematical ideas. Therefore, other students have the opportunity to learn from other colleagues and have access to a variety of different methods in order to solve the same problem.

What could be other characteristics of problems that could help building collective solution spaces those promoting mathematical creativity?

Along with being open-ended, Diezman & Watters (2004) adds that a good math problem should also be complex, contextualized and promote high level of cognitive thinking and problem solving abilities.

Greenes (1997) and Murphy (2004) acknowledge that contextualized and ill-structured problems are important to present to students because they relate more to societies issues. In other words, students face real life situations where they have to find unknown data and use multiple strategies in order to find solutions. Facing unknown or unfamiliar real life situations may bring new and original solutions and strategies. It is also possible to add that real life situations where not all necessary information is known can leave to different interpretations. Hancock (1995) therefore defines problems that can have multiple interpretations as a good math problem. All these types of problems emphasize the importance of being able to develop divergent thinking, a characteristic that Pekhonen (1997) defines as to be a good math problem.

Sheffield (2003) stresses on the importance on giving students investigation problems enabling students to develop different ways of communicating results, letting them develop different mathematical abilities, encouraging students to further explorations once the problem is solved by finding more questions, ideas etc.

As it was shown by Freiman (in the previous chapter), virtual communities represent a new medium to present rich mathematical problems online and thus increase opportunities for more students to communicate their own way of solving the problem, contributing to the creation of collective solution space in which some original solutions can be found and shared.

In the following section, we will analyze this process of potentially creative mathematical work using an example of a (rich and open-ended) mathematical task.

ANALYSIS OF MATHEMATICAL CREATIVITY IN A VIRTUAL PROBLEM SOLVING ENVIRONMENT

The CASMI virtual environment (www.umoncton.ca/casmi) gives opportunities for schoolchildren and university students to solve mathematical problems (Freiman, Lirette-Pitre, Manuel, 2007). The goal of this study was to determine how online problem solving can contribute to the development of mathematical creativity. In our analysis we aim to explore the following research question:

How creative are the solutions of mathematical problems posted by the participants of the CASMI project?

In order to answer this question, we selected a problem that met many of the 'richness' criterions mentioned above. The following text was posted on the CASMI web site (Figure 16.2):

Problem
There exist some superstitions on numbers. For example the number 13 is recognized as bad luck. Therefore, in big cities, buildings don't have a 13th floor.

The number 666 is also known as «mysterious». However, that number is special and needs a particular attention because it can be written as a sum of many consecutive integers. We give you the challenge to find as many possible solutions. Can you find them all?

Figure 16.2 Problem posted on the CASMI web site.

This problem is opened-ended and allows several possible solutions that can be found using multiple strategies. Here are the possible solutions to this problem and possible strategies that can be used.

Solutions to the problem:

- $221 - 223$
- $165 - 168$
- $70 - 78$
- $50 - 61$
- $1 - 36$

Possible strategies:

- Writing mathematical equations like example $x + x + 1 + x + 2 = 666$, $x + x + 1 + x + 2 + x + 3 = 666$ and so on.
- Using number properties like example $666/3 = 222$. So a possible solution would be $222 - 1, 222, 222 + 1$.
- Using arithmetic series
- Trial and error
- Using technologies such as calculator or computers to facilitate search for solutions

This problem is also put in a Halloween-related context. The problem is not ill-structured because all the necessary data and information are given. It can however allow the student to explore the problem further (after having found one solution) by trying to generalize it.

A collective solution space that we considered in our analysis has been built of 139 solutions that were submitted electronically. As characteristics of mathematical creativity, we were looking at fluency, flexibility and originality of correct solutions submitted for a particular mathematical problem. The fluency is defined by the number of correct answers found in the problem. The flexibility will represent the number of different strategies used to find the answers to the problem and solutions that will statistically not be frequent will be defined as original (D. Haylock, 1997; Krutetskii, 1976).

Table 16.1 shows the scoring system for each or the components. The scoring system that we used was inspired from the works of Balka (1974).

TABLE 16.1 Scoring System for Each of the Components Used in the Definition of Mathematical Creativity

Mathematical creativity component	Scoring system
Fluency	• 1 point for each correct answer found • If the solution contained one answer with the sum from 1 to 36 and 0 to 36, it only counted as 1 answer. • No extra points were added if the solution contained few correct answers and ... to represent that there are more possible answers unless that could represent the other answers with a certain relation.
Flexibility	• 1 point per each strategy used to find correct answers. • In order to count 2 different strategies, the student must have said in his solution how he proceeded. Therefore, no points were given if needed to guess the number of strategies in a solution
Originality	• After looking over all the solutions, you give a scoring from 0 to 2 points for each strategy depending on the percentage of people who found the answer or the strategy. • 0 points was given if 20% or more people used that strategy or found that solution • 1 point was given if only between 5% and 20% of people found that answer or used that strategy • 2 points was given if 5% or less people found the answer or used a particular strategy.

Let us look at some striking examples of concrete solutions (translated from French): In each solution, we explained in details why we gave specific scores for each of the criterions of creativity.

The student (Figure 16.3) was able to find 3 correct answers which gave her 3 points for fluency. At the same time, this is one of the most flexible solutions because 3 different strategies were used to find the answers. The first strategy was the use of the balance (+1 −1), after dividing 666 by 3 to

---

The solution (9th grader)

The first sum of consecutive integers that I found was 221, 222 and 223.

I did 666 / 3 = 222. 222 − 1 = 221 and 222 + 1 = 223. Therefore, 221 + 222 + 223 = 666.

I tried using the calculator and started adding each integer and found that adding the numbers from 0 to 36 = 666.

Then I proceeded by trail and error to see if I could find other possibilities and I found integers from 50 to 61 = 666.

I think that there are other solutions, but I haven't been able to find any other so far.

Figure 16.3 A first solution of the problem.

get consecutive numbers. Even though this student didn't use an equation, it remains interesting that she was able to discover the "+1, −1" regularity about consecutive integers. The second strategy that was used was using the calculator and trying adding numbers. The last strategy that was used was trial and error where she found the integers from 50 to 61 as another possibility. We can also see that the student reflected on her solution by saying that she thinks that there are other possible solutions to this problem but she was not able to find them. Not only did she use several cognitive strategies in his problem, but she demonstrated the use of meta-cognitive strategies. This solution obtained a 3 for flexibility. And as for originality, since the first 2 strategies were quite popular and only the last solution was original, she had 2 points for originality.

In this solution (Figure 16.4), the student was able to find all 5 solutions, which gave her a score of 5 on fluency. In this solution, we found 2 different strategies: using the division to find the 3 first integers and using trial and error. This solution scored a 2 in flexibility. As for originality, this solution received a 0 because these 2 strategies were used quite often in the collective solution space. In this solution we do not see any explicit use of meta-cognition. She just admitted that those were the 5 that she found.

In this example (Figure 16.5), the student found 2 different ways to solve the problem, which gave her a score of 2 on flexibility. The 2 strategies used are similar to the ones presented in the previous 2 examples. The first strategy was the use of the balance (+1 −1), after dividing 666 by 3 to get consecutive numbers (Note that he didn't show the balance so we can only assume that it's what he was thinking). The second strategy he used was the use of an algebraic equation, which was considered one of the original strategies (3.9% according to table 4). However, he only found one answer which gave him a score of 1 on fluency. Since one of the 2 strategies was original, the student received a score of 2 for originality.

The solution (8th grader)

HI

I first started by dividing 666 by 3, which gave me 222. So I tried it starting with 222 but it did not work ☹☹ And then I tried starting with 221 and it gave me one answer— 221, 222, 223 = 666. After I went by trial and error. And I found a second solution 70, 71, 72, 73, 74, 75, 76, 77, 78 = 666. After that, I found this answer: 165, 166, 167, 168 = 666. After that i searched for a long time and then I found this answer: 50, 51, 52, 53, 54, 55, 56, 57, 58, 59, 60, 61. And then I found 1 to 36. there are all the solutions that I have found

BYE!! ☺☺

Figure 16.4 A second solution of the problem.

The solution (8th grader)

1st way

3/666 = 222

221 + 222 + 223 = 666

2nd way

X + x + 1 + x + 2 = 666

3x + 3 = 666

3x = 666 − 3

3/3x = 663/3

3/663 = 221

X = 221

Figure 16.5 A third solution of the problem.

RESULTS

Out of total of 139 solutions submitted, 69 of them were not considered because either the answers were incorrect or it was impossible to evaluate the solution because no explanations were given.

The remaining 70 solutions that we analyzed using our criteria had at least one correct answer. Looking at the *fluency*, we found that 37.5% of them had more than one correct answer. Table 16.2 represents the percentage of solutions for each score of fluency (the number of correct answers found).

As for *flexibility*, 11.4% used more than one strategy to find the answers needed (They used 2 strategies and one solution had 3 strategies). Our data show that a mass of students limit their solutions by applying only one strategy with only one strategy to find their answers.

Table 16.3 contains the percentage of solutions that used each strategy in order to solve the problem found in the solutions along with the percentage of solutions that contained each strategy along with the scores on *originality* based on the percentage given in the scoring system. Table 16.3

TABLE 16.2 Percentage of the Number or Correct Answers Found

Number of correct answers (score for fluency)	Percentage (%)
1	62.5
2	10.3
3	12.5
4	8.0
5	5.7

TABLE 16.3 Percentage of Each Strategy Found in the Solutions

Strategy used	Percentage (%)	Originality score
No strategy explained (the solutions only contained the answers)	25.0	0
Trial and error	20.0	0
Number properties (using factors and trying to find regularities)	28.2	0
Using the sums +1 and –1, +2, –2 after dividing 666 by 2, 3, 4 ...	15.6	1
Using technology tools	5.8	1
Algebraic equation	3.9	2

shows that the most original strategies used for this problem were the use of technology (5.8%) or an algebraic equation (3.9%).

DISCUSSION AND CONCLUSIVE REMARKS

The results show that a majority (2/3) of students seem to stop shortly after having found one answer. The explanation given by previous studies that this reflects usual classroom routine (one problem has one answer) seems to be plausible in the case of virtual problem solving. However, as seen, the community has produced a collective solution space that contains some characteristics of fluency because of all the possible answers found.

When it comes to flexibility, only a small percentage of solutions (11.4%) contained traces of different strategies. It is very difficult at this point in our research to assess this characteristic of creativity. But the collective solutions space does contain different strategies. By sharing and discussing them, students may get interested at looking deeper at the problem and trying to find different strategies.

As for the originality, trial and error and the use of number properties or patterns were the most common strategies used by community members (63.8% of all the solutions). The most original (rarely used) strategies used were technology and (surprisingly?) the creation of an algebraic equation (9.7% of solutions submitted). It may be explained on one side by the fact that trial and error strategy may be seen as helpful for students who do not know how to solve the problem at the first look. Using arithmetic before algebra may be due to the age of students (11–13 years old) who lack algebraic skills that are only emerging at this stage (see the New-Brunswick mathematics curriculum) (MENB, 2005).

We presented results of our analysis for just one problem and only as an exploratory study. The data give us little explanation how creativity develops in the virtual space. More sophisticated research tools will be needed to learn it.

Trying to look at what can be done to foster more creative work in mathematics using the web resources like CASMI, we can however make some suggestions based on our results.

First, in the CASMI virtual community, every student that solves a problem gets personal feedback that helps students analyze their solution (Freiman, Manuel, Lirette-Pitre, 2007). We may use this feed-back in order to encourage students to look at different possible answers and strategies.

Second the discussion forum built inside the CASMI virtual community could be used to create debates and discussions on the problems posted on the site as it was .recommend by several authors (Cobb & Beaersfeld, 1995; Lampert & Blunk, 1996; Leikin & Dimur, 2007; Sfard, Nesher, Streefland, Cobb, & Mason, 1998; Steinbring, Sierpinska 1998). By doing so, students could become aware of other (different) strategies that can be possibly used when solving particular problems, thus increasing more of their own creative potential. However, the culture of such online discussion is still to be developed in schoolchildren.

Finally, in the CASMI virtual community, there is a special section where a general comment on the set of solutions is posted. This comment contains analysis of the problem and examples of solutions that were submitted. In order to emphasize originality, I recommend that the examples posted would be the most original solutions that were found during the correction.

We shall acknowledge that presented results are preliminary at this stage of our research. More rigorous research on mathematical creativity in the CASMI virtual community is still underway. This research will study the relationships between the potential creativity of a virtual collective solution space and the richness of the problems. The results of this study could give teachers who use the virtual community to have ideas on the type of problems that have a greater potential of developing mathematical creativity and those that would need more effort in order help students be more creative. Also, our research would suggest the types of problems would help to foster mathematical creativity.

Summarizing our conclusions, the exploratory study seems to demonstrate that a virtual community like CASMI could possibly be an environment that offers opportunities to students to further develop their mathematical creativity by solving open-ended problems using their own strategies and ways of thinking. Also, by having the chance to be able to communicate with members from not only their classroom colleagues but from all over the world via discussion forums, students may get opportunities to learn different ways of thinking and strategies and further their thinking process. Further research is however needed in order to have a better understanding of the impacts of virtual communities on mathematical creativity.

REFERENCES

Anabile, T. M. (1996). *Creativity in context: Update to the social psychology of creativity.* Boulder, CO: Westview Press.

Chan, C. M. E. (2008). *The use of mathematical modeling tasks to develop creativity.* Paper presented at the 11th International Congress on Mathematical Education: Discussion group 9, Monterrey, Mexico (July 6–13 2008).

Cline, S. (1999). *Giftedness Has Many Faces: Multiple Talents and Abilities in the Classroom:* Distributed by Winslow Press for The Foundation for Concepts in Education, 770 East Atlantic Ave., Delray Beach, FL 33483 ($34.95).

Cobb, P., & Beaersfeld, H. (1995). *The emergence of mathematical meaning: Interactions in classroom cultures.* Hillsdale, NJ: Erlbaumm.

Csikszentmihalyi, M. (2000). Implications of a systems perspective for the study of creativity. In R. J. Stemberg (Ed.), *Handbook of creativity* (pp. 313–338). Cambridge UK: Cambridge University Press.

Diezmann, C., Watters, J. . (2004). *Challenge and connectedness in the mathematics classroom: using lateral strategies with gifted elementary students.* . Paper presented at the Topics Study Group 4: Activities and Programs for Gifted Students. The 10th International Congress on Mathematical Education, Copenhagen, Denmark/ Eds.

Feist, G. J. (1998). A meta-analysis of the impact of personnality on scientific and artistic creativity. *Personnality and social psychology review, 2*(4), 90–309.

Freiman, V. (2006). Problems to discover and to boost mathematical talent in early grades: a challenging situations approach. *The Montana Mathematics Enthusiast, 3*(1), 51–75.

Greenes, C. (1997). Honing the Abilities of the Mathematically Promising. *Mathematics Teacher, 90*(7), 582–586.

Gruber, H. E., & Wallace, D. B. (2000). The case study methods and evolving systems approach for understanding unique creative people at work. In R. J. Stemberg (Ed.), *Handbook of creativity* (pp. 93–115). Cambridge, UK: Cambridge University Press.

Hancock, C. L. (1995). Implementing the Assessment Standards for School Mathematics: Enhancing Mathematics Learning with Open-Ended Questions. *Mathematics Teacher, 88*(6), 496–499.

Haylock, D. (1997). Recognizing mathematical creativity in school children. *International Reviews on Mathematical Education, 29*(3), 68–74.

Haylock, D. W. (1987). A framework for assessing mathematical creativity in school children. *Education Studies in Mathematics, 18*(1), 59–74.

Holyoak, K. J., & Thagard, P. (1997). The Analogical Mind. *American Psychologist, 52*(1), 35–44.

Klavir, R., & Hershkovitz, S. (2008). Teaching and evaluating "Open-Ended" Problems [Electronic Version]. *International Journal for Mathematics Teaching and Learning. May 20th. http://www.cimt.plymouth.ac.uk/journal/klavir.pdf,*

Krutetskii, V. A. (1976). *The psychology of mathematical abilities in school children.* Chicago: University of Chicago Press.

Lampert, M., & Blunk, M. l. (1996). *Talking mathematics: Studies of learning and learning.* Cambridge, UK: Cambridge University.

Leikin, R. (2007). *Habits of mind associated with advanced mathematical thinking and solution spaces of mathematical tasks.* Paper presented at the fifth conference of the Europeens society for ressearch in mathematics education-CERME–5.

Leikin, R., & Dimur, S. (2007). Teacher flexibility in mathematical discussion. *Journal of Mathematical Behavior, 26* (2007), 328–347.

Mann, E. (2005). *Mathematical Creativity and School Mathematics: Indicators of Mathematical Creativity in Middle School Students.* University of Connecticut, 2005, www.gifted.uconn.edu/siegle/Dissertations/Eric%20Mann.pdf

Mann, E. (2006). Creativity: The Essence of Mathematics. *Journal for the Education of the Gifted, 30*(2), 236–260.

Meissner, H. (2008). *Intuitive–Creative–Gifted–Logical, an analysis for the discussion group DG at ICME 11.* Paper presented at the 11th International Congress on Mathematical Education, Monterrey, Mexico (July 6–13 2008).

Meissner, H. (2005). *Creativity and mathematics education. .* Paper presented at the The 3rd East Asia Regional Conference on Mathematics Education from http://www.math.ecnu.edu.cn/earcome3/sym1/sym104.pdf

Murphy, E. (2004). Identifying and Measuring Ill-Structured Problem Formulation and Resolution in Online Asynchronous Discussions. *Canadian Journal of Learning and Technology, 30*(1).

NCTM. (2000). *Principles and standards for School Mathematics.* Reston, VA: National Council of Teachers of Mathematics.

Pehkomen, E. (1997). The state-of-art in mathematical creativity. *Internation Reviews on Mathematical Education, 29,* 63–66.

Renzulli, J. S. (1998). The Three-Ring conception of giftedness. In S. Baum, S. M. Reis & L. R. Maxfield (Eds.), *Nurturing the gifts and talents of primary grade students.* Mansfield Center, CT: Creative learning press.

Runco, M. A. (1993). Divergent thinking, creativity and giftedness. *Gifted child Quarterly, 37*(1), 16–22.

Sfard, A., Nesher, P., Streefland, L., Cobb, P., & Mason, J. (1998). Learning mathematics through conversation: Is it as good as they say? *For the learning of mathematics, 18*(1), 41–51.

Sheffield, L. (2003). *Extending the challenge in Mathematics.* California: TAGT & Corwin Press.

Sheffield, L. (2008). *Promoting creativity for all students in mathematics education: An overview.* Paper presented at the 11th International Congress on Mathematical Education: Discussion group 9, Monterrey, Mexico (July 6–13 2008).

Singh, B. (1988). *Teaching-learning strategies and mathematical creativity.* Delhi, India: Mittal Publications.

Sriraman, B. (2004). The characteristics of mathematical creativity. . *The Mathematics Educator, 14*(1), 19–34.

Sriraman, B, & Freiman, V. (2007). Does mathematics gifted education need a philosophy of creativity? *Mediterranean Journal for Research in Mathematics Education, 6*(1 & 2), 23–46.

Steinbring, H., Sierpinska, A., M. G. B.-B. (Eds.). (1998). *Language and communication in the mathematics classroom.* Reston, VA: NCTM.

Stemberg, R. J., & Lubart, T. I. (1996). Investing in creativity. *American Psychologist, 51*, 677–688.

Stemberg, R. J., & Lubart, T. I. (1999). The concept of creativity: Prospects and paradigmes. In R. J. Stemberg (Ed.), *Handbook of creativity.* Cambridge, UK: Cambridge University Press.

Takahashi, A. (2000). Open-ended Problem Solving Enriched by the Internet. NCTM annual meeting: http://www.mste.uiuc.edu/users/aki/open_ended/.

Treffinger, D. J., & Isaksen, S. G. (2005). Creative Problem Solving: The History, Development, and Implications for Gifted Education and Talent Development. *Gifted Child Quarterly, 49*(4), 342.